Software Engineering and Data Science

Software Engineering and Data Science

Editor

Davide Tosi

MDPI • Basel • Beijing • Wuhan • Barcelona • Belgrade • Manchester • Tokyo • Cluj • Tianjin

Editor
Davide Tosi
University of Insubria
Italy

Editorial Office
MDPI
St. Alban-Anlage 66
4052 Basel, Switzerland

This is a reprint of articles from the Special Issue published online in the open access journal *Future Internet* (ISSN 1999-5903) (available at: https://www.mdpi.com/journal/futureinternet/special_issues/SE_DS).

For citation purposes, cite each article independently as indicated on the article page online and as indicated below:

LastName, A.A.; LastName, B.B.; LastName, C.C. Article Title. *Journal Name* **Year**, *Volume Number*, Page Range.

ISBN 978-3-0365-6440-1 (Hbk)
ISBN 978-3-0365-6441-8 (PDF)

Cover image courtesy of Davide Tosi

Contents

About the Editor

Davide Tosi

Davide Tosi is an Associate Professor and Dean of Computer Science degree at the University of Insubria and an Adjunct Professor at the University of Bocconi. His research interests range from big data to software quality.

 *future internet*

MDPI

Editorial

Editorial for the Special Issue on "Software Engineering and Data Science"

Davide Tosi

Department of Theoretical and Applied Sciences (DiSTA), Università degli Studi dell'Insubria, 21100 Varese, Italy; davide.tosi@uninsubria.it

Citation: Tosi, D. Editorial for the Special Issue on "Software Engineering and Data Science". *Future Internet* **2022**, *14*, 306. https://doi.org/10.3390/fi14110306

Received: 24 October 2022
Accepted: 24 October 2022
Published: 26 October 2022

Publisher's Note: MDPI stays neutral with regard to jurisdictional claims in published maps and institutional affiliations.

In the last few years, data-driven software solutions have attracted a lot of attention in research and development at academic, industry, business, and government levels to exploit the hidden knowledge and big data that can be offered to cities and citizens in the future. However, data-driven software solutions are different from "traditional" software development projects, as the focus of the main development core is on managing data (e.g., data store and data quality) and designing behavioral models with the aid of artificial intelligence and machine learning techniques. To this end, new life-cycles, algorithms, methods, processes, and tools are required. The Special Issue, "Software Engineering and Data Science", in the Journal of Future Internet, is devoted to recent trends and advancements in the field of engineering data-intensive software solutions to address challenges in developing, testing, and maintaining such data-driven systems. We received 13 submissions; after the initial screening and the peer review process, six papers have been finally accepted for publication. Accepted articles can be classified into two sets: (1) application of data-driven solutions to real-life problems and (2) techniques and algorithms addressing the different challenges of data-driven software engineering.

The first set of articles discusses the applicability of data science and data-driven solutions to everyday problems. Casini et al. [1] studied the inversion in the decreased/increased rate of new SARS-COV-2 infections in the countries involved in the European football championship that took place from 11 June to 11 July 2021, investigating the hypothesis of an association. They collected and analyzed all data regarding COVID-19 infections from the official online repositories. Then, they adopted Bayesian piecewise regression with a Poisson generalized linear model to look for changepoints in the time series of the new SARS-COV-2 cases of each country involved in the 2020 European football championship. For all the 17 countries involved, the changepoint coincides with an inversion in the SARS-COV-2 case rate from a decreasing to an increasing rate of infections, thus suggesting an association between infection rates and the European football championship. Another example of applying data science to real-life is presented in the work of Tosi et al. [2]. They conducted a correlation study using heterogeneous data sources, such as Google mobility data, SARS-COV-2 infection data, and the official dataset relating to infections in Italian schools for the period of 14 September 2020–30 October 2020. Three extensive Italian regions (Lombardy, Campania, and Emilia) (that adopted different approaches in opening and closing schools to contrast infections) have been deeply studied to understand the main driver that sparked the second SARS-COV-2 wave in Italy. The conducted data analyses suggest that schools are a driver of contagion and are not a safe environment by definition. Munjal et al. [3] applied big data-driven solutions to smart cities. Smart cities will be equipped with millions of smart devices and network connections, thus requiring a high level of energy consumption and carbon emissions. The authors defined a public transport-assisted data-dissemination system to utilize public transport as another communication medium, along with other networks, with the help of software-defined technology. The main objective is to minimize energy consumption with maximum data delivery. To this end, a multi-attribute decision-making algorithm is designed to self-identify the best network among wired, wireless, and public transport networks based on users' requirements and different services. Once public transport was selected as the best network, the Capacitated Vehicle Routing Problem (CVRP) will be implemented to offload data onto buses as per the maximum capacity of buses.

The second set of articles discusses new development methodologies, algorithms for software libraries recommendation, and technologies for ontology-based knowledge extraction from various heterogeneous sources. Almedia et al. [4] addressed the combined adoption of Agile and DevOps software development methodologies to cope with the increasing complexity of managing customer requirements and development requests. The authors presented a qualitative methodology to analyze the benefits that can arise from the combination of the two methodologies. A comprehensive set of twelve case studies, representing practices of the simultaneous adoption of both methodologies, was assessed. The simultaneous adoption of Agile and DevOps, when properly combined and aligned, allows (1) developers to gain greater control over the environment, infrastructure, and applications; (2) a more collaborative and Agile framework; (3) to simplify and automate the model processes to make them more rational and efficient. Krasanakis et al. [5] studied how to help developers automatically discover libraries to be reused in their software projects. They extended the accurate project–library recommendation systems, which employ Graph Neural Networks, with a revised collaborative graph filtering mechanism. The revised filtering mechanism exploits partially absorbing random walk filters, which the authors theorized could emulate human-driven library discovery. The experimental results on a real-world dependency graph of Android project third-party library dependencies highlighted promising research directions in automated software engineering and broader collaborative filtering research. Sikelis et al. [6] provided insight into critical aspects of ontology-based knowledge extraction from various heterogeneous sources, such as text, databases, and human expertise, realized in feature selection. Ontology-based algorithms and approaches are described to represent features and perform feature selection and classification. Moreover, the authors highlighted open issues and challenges related to the research topic of ontology-based knowledge extraction.

We would like to thank all the authors for the papers they submitted to this Special Issue. We would also like to acknowledge all the reviewers for their careful and timely reviews which helped to improve the quality of this Special Issue.

Funding: This research received no external funding.

Conflicts of Interest: The authors declare no conflict of interest.

References

1. Casini, L.; Roccetti, M. A bayesian analysis of the inversion of the SARS-COV-2 case rate in the countries of the 2020 European Football Championship. *Future Internet* **2021**, *13*, 8. [CrossRef]
2. Tosi, D.; Campi, A.S. How schools affected the COVID-19 pandemic in Italy: Data analysis for Lombardy Region, Campania Region, and Emilia Region. *Future Internet* **2021**, *13*, 5. [CrossRef]
3. Munjal, R.; Liu, W.; Li, X.; Gutierrez, J.; Chong, P.H.J. Multi-Attribute decision making for energy-efficient public transport network selection in smart cities. *Future Internet* **2022**, *14*, 42. [CrossRef]
4. Almeida, F.; Simões, J.; Lopes, S. Exploring the benefits of combining DevOps and Agile. *Future Internet* **2022**, *14*, 63. [CrossRef]
5. Krasanakis, E.; Symeonidis, A. Fast library recommendation in software dependency graphs with symmetric partially absorbing random walks. *Future Internet* **2022**, *14*, 124. [CrossRef]
6. Sikelis, K.; Tsekouras, G.E.; Kotis, K. Ontology-based feature selection: A survey. *Future Internet* **2021**, *13*, 158. [CrossRef]

future internet

Article

A Bayesian Analysis of the Inversion of the SARS-COV-2 Case Rate in the Countries of the 2020 European Football Championship

Luca Casini and Marco Roccetti *

Department of Computer Science and Engineering, University of Bologna, 40127 Bologna, Italy; luca.casini7@unibo.it
* Correspondence: marco.roccetti@unibo.it

Abstract: While Europe was beginning to deal with the resurgence of COVID-19 due to the Delta variant, the European football championship took place from 11 June to 11 July 2021. We studied the inversion in the decreased/increased rate of new SARS-COV-2 infections in the countries of the tournament, investigating the hypothesis of an association. Using a Bayesian piecewise regression with a Poisson generalized linear model, we looked for a changepoint in the timeseries of the new SARS-COV-2 cases of each country, expecting it to appear not later than two to three weeks after the date of their first match. The two slopes, before and after the changepoint, were used to discuss the reversal from a decreasing to an increasing rate of the infections. For 17 out of 22 countries (77%) the changepoint came on average 14.97 days after their first match (95% CI 12.29–17.47). For all those 17 countries, the changepoint coincides with an inversion from a decreasing to an increasing rate of the infections. Before the changepoint, the new cases were decreasing, halving on average every 18.07 days (95% CI 11.81–29.42). After the changepoint, the cases begin to increase, doubling every 29.10 days (95% CI 14.12–9.78). This inversion in the SARS-COV-2 case rate, which happened during the tournament, provides evidence in favor of a relationship.

Keywords: SARS-COV-2; Bayesian regression; changepoint detection; European football championship

Citation: Casini, L.; Roccetti, M. A Bayesian Analysis of the Inversion of the SARS-COV-2 Case Rate in the Countries of the 2020 European Football Championship. *Future Internet* **2021**, *13*, 212. https://doi.org/10.3390/fi13080212

Academic Editor: Davide Tosi

Received: 2 August 2021
Accepted: 15 August 2021
Published: 17 August 2021

1. Introduction

Europe, as well as other countries around the world, is seeing a resurgence in the COVID-19 pandemic, after a brief respite given by the effects of the vaccination that started in the first half of 2021. This new wave of the pandemic seems to be driven by a new strain of virus that has been referred to as the Delta variant. This is the scenario in which the European football championship has taken place, from 11 June to 11 July 2021 (one year later than it should have been). This 2020 edition, being a special celebration for the 60th anniversary of the tournament, has had the peculiarity of being hosted by several different countries, instead of just one as it normally happens.

The decision to allow such a massive event across the European continent, in such a delicate time, immediately triggered a debate on the problems it would cause. Nonetheless, the competition was held, leaving each hosting country some freedom on which restrictions to apply (e.g., the number of fans allowed at each football stadium). This resulted in very different behaviors, ranging from Hungary hosting its matches at full stadium capacity at Puskás Arena (~68 thousand seats) to Germany limiting the attendance to 22% of the maximum stadium capacity [1–4]. Obviously, there were more factors than just the stadium, with fans, massively gathering in pubs, squares, and public places, to watch the matches, thus leading to infection clusters that surged all around Europe, as witnessed by the media coverage of these events [5–8]. Not only that but even the gathering of teams and their staff may have given their contribution to the spread of the virus (given the itinerant nature of this edition), as the COVID-19 literature on football and other sports suggests [9–11].

On one side, one could conclude that those who considered this event to be a minor risk did not take into any consideration of those theories that maintain that, with COVID-19, super-spreading events may be the main driver of an epidemic spread, under specific circumstances [12,13]. An example, on 19 February 2020, was the Champions League match, between Atalanta and Valencia, which attracted a third of Bergamo's population to Milan's San Siro stadium. In addition, more than two thousand and a half of Spanish supporters took part. Experts, now, point to that 2020 football match as one of most relevant reasons why the city of Bergamo had become the epicenter of the COVID-19 pandemic, during the first wave in Italy, with a very high death toll; not to mention, that the 35% of Valencia's team also became infected [14]. On the other hand, it is well known that the return of supporters to stadiums is the highest priority for football's business, and the financial impact of the COVID-19 pandemic on football depends, almost exclusively, on both the timing and the scale of supporters' return to stadiums [15].

Following this debate, this work focused on the European football championship and its matches, looking for a possible compatibility with the reversal of the decrease/increase trend of the SARS-COV-2 cases, observed in many countries participating in the tournament. To investigate the hypothesis of an association between those football matches and the resurgence of the virus, we searched for a changepoint in the daily timeseries of the new SARS-COV-2 cases registered in each country, expecting it to appear not later than 2–3 weeks after the date of the first match that the national team played. Upon finding such a changepoint, we investigated if that changepoint was coincidental with a change in the infection rate, from a decreasing trend to an increasing one. It should be noted that our type of analysis has been observational in nature, and it was used to determine if the exposure to the specific risk factor, given the frequent mass gatherings following the football events, might have correlated with the particular outcome of the virus resurgence in many European countries. With this type of study, we cannot demonstrate any cause and effect, but we can make preliminary inferences on the correlation between the participation in the European football championship of a given country and the inversion in the SARS-COV-2 case rate that may have hit, at a particular point in time, the population living in that country.

We can anticipate that 17 out of 22 countries (77%) had a reversal from a decreasing to an increasing rate of the infections, which is temporally coincident with their participation in the European football championship, thus providing evidence to the hypothesis of a link between the upturn of new cases and the tournament. Instead, only 4 out of 12 countries (33%) that did not take part in the tournament (subject of an additional investigation) followed the same pattern as above. This further confirms that, while it can be inferred that an increase in COVID-19 cases may have been an inevitable consequence of the general European situation in July 2021, the European football tournament, with its mass gatherings, has at least played an important role of the accelerator of this phenomenon for many of its participating countries.

The remainder of the paper is structured as follows: In the next section, we describe more precisely the data we used, their sources, and the methodologies we employed. Section 3 presents the results we obtained, while Section 4 discusses them, along with their limitations, and concludes the paper, presenting our final considerations.

2. Materials and Methods

In this section, we provide a description of the data on which our analysis is based, along with the methods used for its collection and the sources from which we collected them (Section 2.1). Then, we present the methodologies we have chosen to conduct our analysis (Section 2.2).

2.1. Data Collection

The timeframe for this study starts two weeks before the start of the tournament on 28 May, and it ends two weeks after the final match on 25 July 2021. All data regarding

COVID-19 infections were collected from the online repository: Our World in Data [16], that in turn aggregates various sources. In particular, the confirmed cases were provided by the COVID-19 Data Repository by the Center for Systems Science and Engineering at the Johns Hopkins University. The timeseries of daily confirmed cases was then smoothed using a rolling average with a 7 day-long window. This was useful for removing the periodicity patterns of the various testing and registering case processes, with some countries that unfortunately release numbers once every few days (e.g., Sweden) or slow down on weekends (e.g., Italy).

Data for the European football championship were collected from the relevant Wikipedia page [17]. We looked at the participating countries, their first and last matches in the competition, and their last hosted match (if they were a hosting country). These dates were then compared with the changepoints found with the Bayesian method described in the next section. For the sake of simplicity, given that the data for the United Kingdom were given as a whole in the dataset we used, we considered Wales, Scotland, and England as a single entity, even if the three countries participated individually.

We conclude this subsection by confirming that patients and/or the public were not involved in the design, conduct, reporting, or dissemination plans of this research. All data come from a publicly available repository where they are stored in an aggregated and anonymized format.

2.2. Bayesian Changepoint Detection and Analysis

Using a changepoint estimation technique, based on a Bayesian piecewise regression, we have looked for a changepoint in the trend of the infection curve, whether it was growing or falling. In particular, we fitted a Poisson generalized linear model where the dependent variable was the number of new daily confirmed SARS-COV-2 cases, and the independent variable was just the number of days since 28 May 2021 (until 25 July). The result was a model comprised of a changepoint and two segments, whose slopes represent, respectively, the increase/decrease in case rate before the changepoint and the increase/decrease in case rate after it. The fact that our interest was not in modeling the spread of the virus with the maximum precision but rather in finding the point in time when the infection rate inverted (or simply changed) its trend, with the added bonus of a Bayesian uncertainty estimation, is worth noting. The model takes the mathematical form below:

$$\ln(E(Y|x)) = a_1 + xb_1 \text{ if } x < \tau \qquad \ln(E(Y|x)) = a_2 + xb_2 \text{ if } x > \tau \tag{1}$$

It is worth noting that our dependent variable (the confirmed daily cases) was modelled as a Poisson distribution, whose mean depends on the regression coefficients a_1 and b_1, respectively, along with the intercept and angular coefficient before the changepoint τ (while a_2 and b_2 play the same role after the changepoint). To be considered are the three following facts:

First, since the two regression lines are joined at the changepoint τ; the second intercept term a_2 is not estimated as it is bound to be $a_2 = \tau (b_1 - b_2) + a_1$.

Second, the formula above returns the exponential growth/decay trend, both before and after τ, as easily identifiable slopes.

Third, to compute the number of days needed to halve/double the number of cases before/after a changepoint, the following two formulas can be used: specifically, Formula (2) can be used to compute the halving (H) and doubling (D) time, before a changepoint:

$$H_b = \frac{\ln \frac{y_\tau}{2} - a_1}{b_1} - \tau \, ; \, D_b = \frac{\ln 2y_\tau - a_1}{b_1} - \tau \, ; \, \text{where } y_\tau = e^{a_1 + \tau b_1} \tag{2}$$

Formula (3), instead, can be used to compute the halving and doubling time, after a changepoint.

$$H_a = \frac{\ln \frac{y_\tau}{2} - \ln y_\tau}{b_2} \; ; \; D_a = \frac{\ln 2y_\tau - \ln y_\tau}{b_2} \; ; \text{ where } y_\tau = e^{a_1 + \tau b_1} \tag{3}$$

To fit the model above, we used the R package mcp, using a Markov chain Monte Carlo method [18]. For starting the Bayesian estimation, the default priors for τ, a_1, a_2, and b_1 were chosen as suggested in [18], thus considering the prior of τ as a uniform, and the parameters a_1, b_1, and b_2 as normally distributed, as reported in the following formulas:

$$\tau \sim \text{Uniform } (\min(x), \max(x)) \tag{4}$$

$$a_1, b_1, b_2 \sim \mathcal{N}(0, 10) \tag{5}$$

It is now worth noting that the mean value of the computed changepoint posterior distribution was used to calculate the distance in time between the date of a given changepoint and that of the first match played the corresponding team. Similarly, the mean values for the coefficients b_1 and b_2 were used to compute the steepness of the two slopes, respectively, before and after the changepoint.

The values obtained from the Bayesian regression have 95% credible intervals, associated with them. The aggregated statistics we computed for the countries (average distance from changepoint, average doubling/halving time, etc.) have 95% confidence intervals, computed using bootstrap.

This completes the description of our method from a statistical viewpoint. Nonetheless, it is appropriate to motivate the reason behind the use of this statistical methodology. The intuition is as follows: we were interested in finding if there was a particular point in time (occurring during the championship) that had brought a change in the curve of the number of the new daily infections, something like: a before and an after. In such a case, we also wanted to have some clear mathematical representations describing the increase or the decrease in the number of cases, what we could call the growth/decay rates.

We have obtained this by fitting a regression model that is segmented (i.e., piecewise). The precise point of the change was found by looking for the place that yielded the best fit with the regression. Not only that, we have also chosen to use a Bayesian regression, as it makes the model more interpretable, especially in the case of a bad fit (e.g., multiple changepoints, when we look for just one).

At that point, once we have obtained our posterior distribution on the parameters of interest, we have then used the mean value of the changepoint distribution to compare it with the date of the first match that each given national team had played, to see the existence of some relationship. Here, the idea is that if: (i) no more than two or three weeks separate these two events (first match and changepoint), and (ii) the change returns an inversion in the infection rate from a decrease to an increase trend, then we can strengthen the suspicion that the tournament with its mass gatherings played the role of the accelerator of a broader infection increase trend in Europe.

To complete this informal description, it is worth mentioning that we have used the mean values of the parameters from the distribution to draw the two straight lines representing the rate of the new cases before and after the changepoint, and finally, we have used them to compute the number of days needed to double/halve the number of cases. This final computation gives one a more precise idea of the impact of the change.

As a final note, it is important to mention that while it is quite common that COVID-19 cases show their biggest single-day jumps two to three weeks after a particular mass event [19], we have extended the search space for a changepoint to four weeks, for the sake of reliability. Nonetheless, following the literature, we have considered to be of interest only those changes that occurred in the infection curves in the temporal interval from 5–6 to 22–23 days after the event of interest.

3. Results

This section is split into two different parts. The first one (Section 3.1) reports the results we obtained with the 22 countries that took part in the European Football Championship. The second one (Section 3.2) illustrates the results we obtained with some 12 European countries that did not participate in the tournament.

3.1. Countries That Participated in the Tournament

In total, 17 out of 22 (77%) countries taking part in the European football championship show a changepoint occurring not later than 2–3 weeks after their first match (i.e., during the tournament).

For all these 17 countries, the changepoint coincides with a reversal in the new daily SARS-COV-2 cases, from a decreasing to an increasing rate.

The group of all these countries provides an evidence in favor of the hypothesis. Precisely, the group is comprised of all the following countries: Austria, Belgium, Croatia, Czechia, Denmark, Finland. France, Germany, Hungary, Italy, Netherlands, North Macedonia, Poland, Slovakia, Spain, Switzerland, and Ukraine.

Table 1 provides the lists of those countries, where under the τ we listed, for each country, the mean value of the days passed before the changepoint was detected since 28 May 2021 (i.e., the beginning of the period of observation). Since we are working with a posterior distribution, the 95% CI is indicated in brackets.

Table 1. Countries with a changepoint coincidental with a reversal from a decrease to an increase in the SARS-COV-2 case rate that occurred during the European football championship.

Country (Participating in the Tournament)	τ (Changepoint, avg. Value and 95% CI)	Diff (Days Separating τ from First Match)	b_1 (Angular Coefficient before τ, avg. Value and 95% CI)	b_2 (Angular Coefficient after τ, avg. Value and 95% CI)	a_1 (Intercept before τ, avg. Value and 95% CI)
Austria	36.4 (35.8, 37.1)	20	−0.05 (−0.06, −0.05)	0.08 (0.08, 0.09)	6.28 (6.25, 6.32)
Belgium	24.9 (24.6, 25.2)	10	−0.06 (−0.06, −0.06)	0.04 (0.04, 0.04)	7.70 (7.68, 7.71)
Croatia	28.7 (27.4, 30.3)	13	−0.06 (−0.06, −0.06)	0.02 (0.02, 0.03)	5.87 (5.83, 5.92)
Czechia	26.2 (25.3, 27.2)	9	−0.06 (−0.06, −0.05)	0.02 (0.02, 0.03)	6.30 (6.26, 6.34)
Denmark	28.2 (27.8, 28.6)	13	−0.06 (−0.06, −0.06)	0.05 (0.05, 0.06)	7.13 (7.11, 7.15)
Finland	19.8 (18.4, 21.0)	5	−0.03 (−0.04, −0.03)	0.04 (0.04, 0.05)	4.98 (4.90, 5.05)
France	34.7 (34.6, 34.8)	17	−0.06 (−0.06, −0.06)	0.11 (0.11, 0.11)	9.28 (9.28, 9.29)
Germany	35.1 (34.6, 35.5)	17	−0.07 (−0.07, −0.07)	0.05 (0.05, 0.05)	8.60 (8.59, 8.62)
Hungary	40.3 (38.4, 42.0)	22	−0.06 (−0.06, −0.06)	0.03 (0.02, 0.05)	5.99 (5.95, 6.03)
Italy	36.5 (36.3, 36.8)	23	−0.05 (−0.05, −0.05)	0.09 (0.09, 0.10)	8.25 (8.24, 8.26)
Netherlands	26.4 (26.2, 26.6)	10	−0.06 (−0.06, −0.06)	0.09 (0.09, 0.09)	8.18 (8.17, 8.20)
N. Macedonia	34.8 (31.9, 37.6)	19	−0.05 (−0.05, −0.04)	0.05 (0.04, 0.07)	3.59 (3.46, 3.72)
Poland	35.2 (33.5, 36.8)	18	−0.07 (−0.07, −0.07)	0.01 (−0.00, 0.01)	6.92 (6.89, 6.94)
Slovakia	39.4 (36.8, 42.1)	22	−0.05 (−0.05, −0.04)	0.02 (0.00, 0.03)	5.02 (4.96, 5.08)
Spain	24.9 (24.8, 25.0)	8	−0.01 (−0.01, −0.01)	0.07 (0.06, 0.07)	8.45 (8.44, 8.46)
Switzerland	32.9 (32.5, 33.5)	18	−0.07 (−0.07, −0.07)	0.08 (0.08, 0.08)	6.93 (6.91, 6.96)
Ukraine	25.8 (25.1, 26.5)	10	−0.05 (−0.05, −0.05)	0.00 (0.00, 0.01)	8.06 (8.04, 8.07)

In the *diff* column, instead, we listed the difference, in terms of days, between the point in time when the changepoint occurred and the date of the first match played by that given national team.

The fourth and fifth columns of Table 1 show the mean values (with the corresponding 95% CI) for the coefficients b_1 and b_2, that have been used to compute the steepness of the slopes, respectively, before and after the changepoint.

The sixth column, finally, reports the average value of the first intercept a_1, with its 95% CI.

We further worked with the numbers comprised in Table 1 by rounding the mean changepoint value for all the 17 countries and then by calculating the difference, in terms of days, between that value and the date when they played their first match.

This way, we obtained that the average date of the changepoint, for all the 17 countries of interest, falls 14.97 days (95% CI 12.29–17.47) after the beginning of their participation in the tournament (approximately two weeks).

Finally, we made a step further and, taking the mean values for the coefficients b_1 and b_2, we estimated how the slopes for the two lines changed, on average, before and after the changepoint. We gathered that all the 17 countries had a decreasing number of daily cases until the changepoint and ended up with a reversed trend afterwards.

Table 2 shows the halving time before, and the doubling time, after the changepoint, for each given country of this group.

Table 2. Quantifying the inversion from a decrease to an increase in the SARS-COV-2 case rate for the countries of Table 1.

Country	Days Needed to Halve the Number of Cases (before τ)	Days Needed to Double the Number of Cases (after τ)
Austria	12.69	8.18
Belgium	11.05	17.60
Croatia	11.77	28.32
Czechia	12.05	28.22
Denmark	11.49	12.71
Finland	21.81	15.84
France	11.86	6.32
Germany	10.14	13.17
Hungary	11.10	22.10
Italy	13.56	7.43
Netherlands	12.11	7.69
N. Maced.	14.88	13.20
Poland	9.67	92.80
Slovakia	14.91	41.59
Spain	103.50	10.62
Switzerland	10.03	8.56
Ukraine	14.43	159.00

More precisely, the mean halving time before the changepoint is 18.07 days (95% CI 11.81–29.42), while the mean doubling time after the changepoint is 29.10 days (95% CI 14.12–49.78).

The credible intervals are quite wide, but if we better investigate the values reported in Table 2, we recognize that most of the deviation depends on just three countries, namely: Spain, Ukraine, and Poland, with their exceptionally large values.

To better highlight and summarize all the results we have discussed so far, we also present Figures 1 and 2, where the same results are portrayed from a clear graphical viewpoint.

In particular, Figure 1 takes into account the inversion of the SARS-COV-2 case trend of the following countries: Austria, Belgium, Croatia, Czechia, Denmark, Finland, France, Germany, Hungary, and Italy.

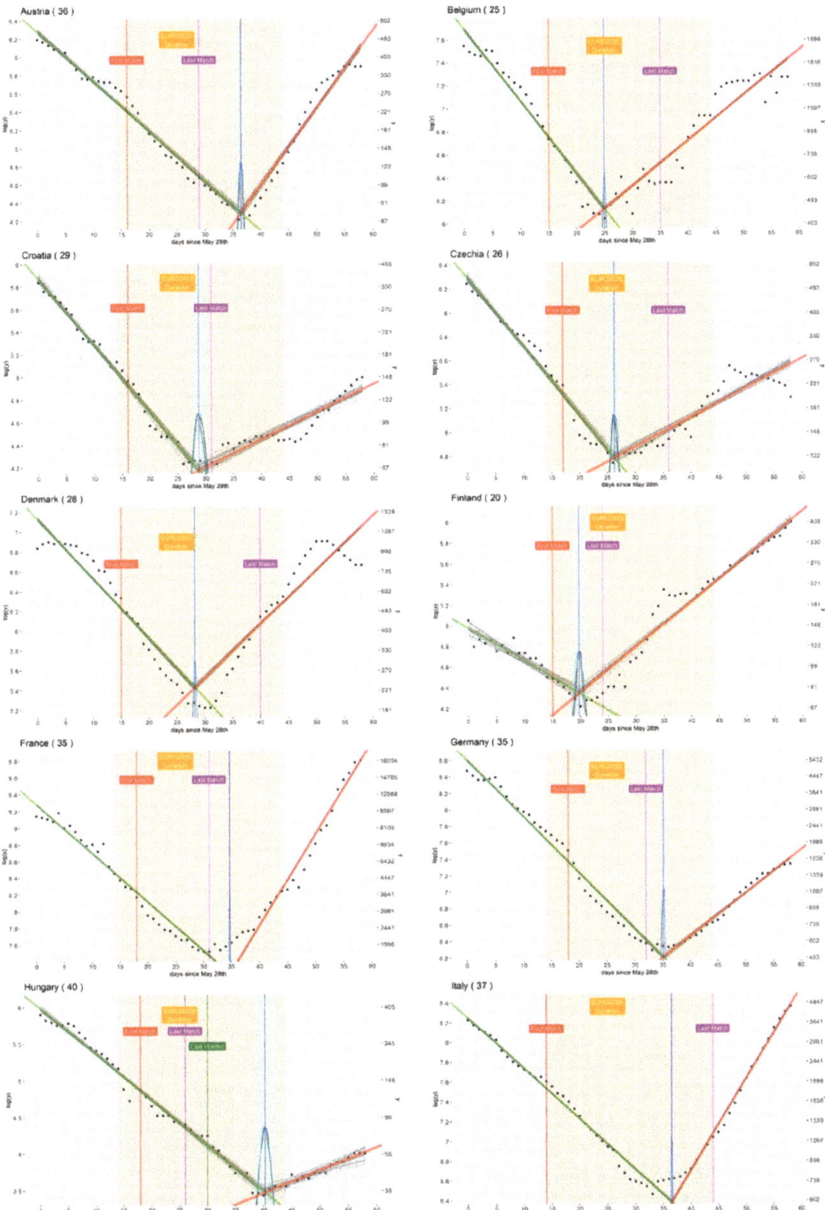

Figure 1. Inversion of the SARS-COV-2 case trend for Austria, Belgium, Croatia, Czechia, Denmark, Finland. France, Germany, Hungary, and Italy, occurring not later than 2–3 weeks after their first match. Yellow space: duration of the tournament. Red vertical line: first match. Purple vertical line: last match. Green vertical line: last hosted match. Blue vertical line: changepoint. Blue space: CI amplitude for the changepoint. Blue bell-shaped peaks: peaks of the probability density function for the changepoint. Green segment: case rate trend before the changepoint. Red segment: case rate trend after the changepoint. Grey segments: fitted lines drawn randomly from the posterior distribution, based on the corresponding CI. Black dots: number of daily CARS-COV-2 cases. Rightmost y axis: number of cases. Leftmost y axis: logarithm of the number of cases.

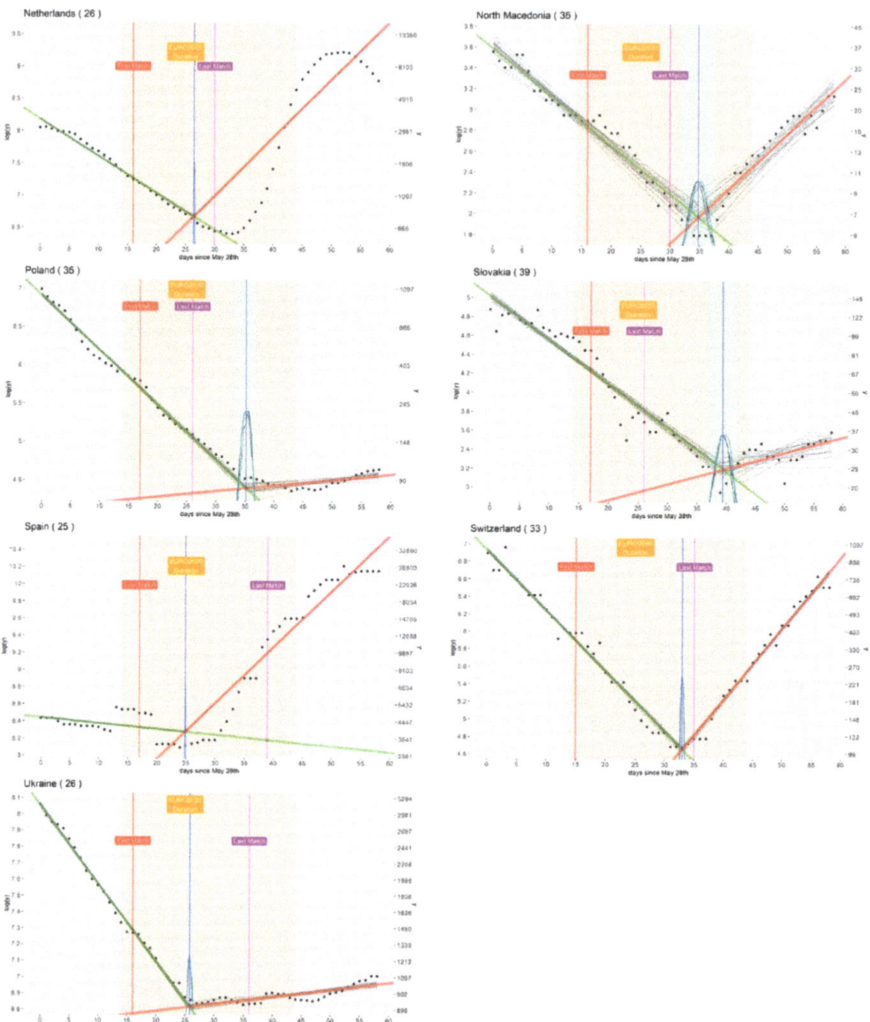

Figure 2. Inversion of the SARS-COV-2 case trend for the Netherlands, North Macedonia, Poland, Slovakia, Spain, Switzerland, and Ukraine, occurring not later than 2–3 weeks after their first match. Yellow space: duration of the tournament. Red vertical line: first match. Purple vertical line: last match. Green vertical line: last hosted match. Blue vertical line: changepoint. Blue space: CI amplitude for the changepoint. Blue bell-shaped peaks: peaks of the probability density function for the changepoint. Green segment: case rate trend before the changepoint. Red segment: case rate trend after the changepoint. Grey segments: fitted lines drawn randomly from the posterior distribution, based on the corresponding CI. Black dots: number of daily CARS-COV-2 cases. Rightmost y axis: number of cases. Leftmost y axis: logarithm of the number of cases.

Figure 2, instead, shows the inversion of the SARS-COV-2 case trend of the Netherlands, North Macedonia, Poland, Slovakia, Spain, Switzerland, and Ukraine. All the relevant information needed to interpret the two figures was inserted in the corresponding captions.

We used two separate figures, just for the sake of manageability. At the end, also based on an analysis of these figures, we can maintain that these results are fully compatible with the tournament being a factor.

All the five remaining countries (i.e., Portugal, Russia, Sweden, Turkey, and the UK), instead, break the pattern and cannot be considered an evidence in favor of the research hypothesis. In particular: (i) Portugal, Russia, and the UK show a robust increasing trend in the SARS-COV-2 infection case, starting well before the beginning the tournament; hence, the detected changepoints, as well as the relative slopes, cannot considered to be evidence in favor the hypothesis; (ii) Turkey seems to show quite a regular pattern, with a well identifiable changepoint and the usual inverting trend in the case rate; nonetheless, the problem is that that changepoint happens well after the team left the competition, more than four weeks since its first match; and (iii) finally, for Sweden, the model fails to fit because there seem to be two different changepoints, that are either before or after the tournament, making them irrelevant. The situations mentioned above are illustrated in Figure 3, where it is evident that all those five countries break the pattern. Again, all the relevant information needed to interpret Figure 3 was inserted in the corresponding caption.

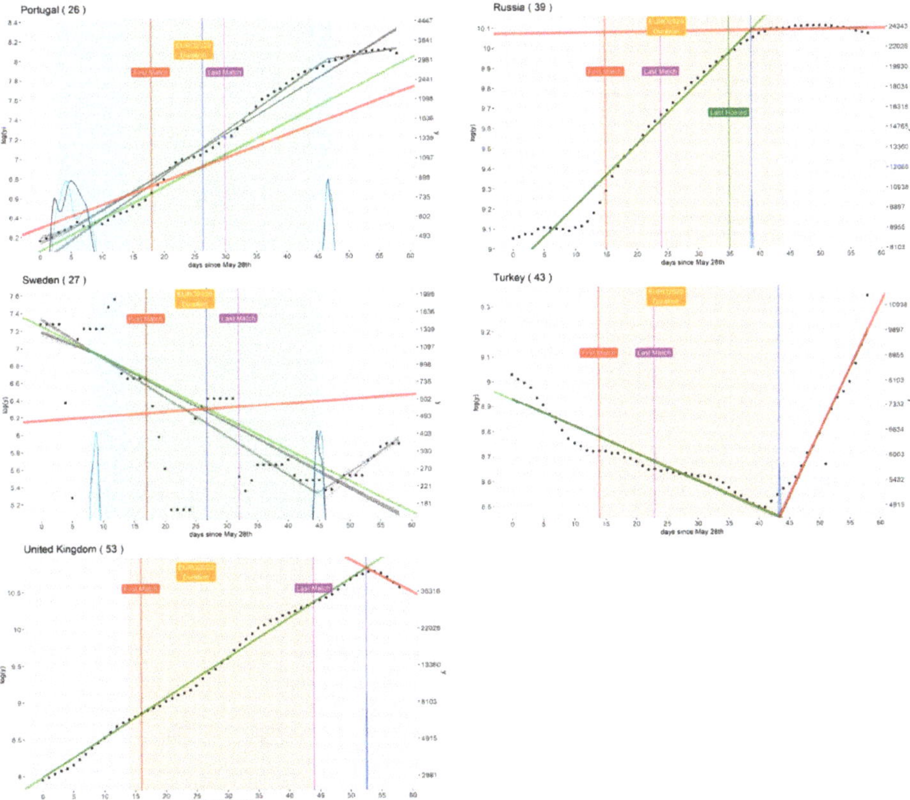

Figure 3. Portugal, Russia, Sweden, Turkey, and the UK break the pattern, without: (i) a well-recognizable changepoint and (ii) a reversal from a decrease to an increase in the SARS-COV-2 case rate, occurring not later than 2–3 weeks after the beginning of the tournament. Yellow space: duration of the tournament. Red vertical line: first match. Purple vertical line: last match. Green vertical line: last hosted match. Blue vertical line: changepoint. Blue space: CI amplitude for the changepoint. Blue bell-shaped peaks: peaks of the probability density function for the changepoint. Green segment: case rate trend before the changepoint. Red segment: case rate trend after the changepoint. Grey segments: fitted lines drawn randomly from the posterior distribution, based on the corresponding CI. Black dots: number of daily CARS-COV-2 cases. Rightmost *y* axis: number of cases. Leftmost *y* axis: logarithm of the number of cases.

Finally, Table 3 reports the value of τ, *diff*, and of all the other parameters, with the corresponding 95% CI. Of particular interest, here, is the large excursion in the CIs for Sweden and Portugal that witnesses the peculiarity of that situation.

Table 3. Regression parameters for the five countries that break the pattern.

Country (Participating in the Tournament)	τ (Changepoint, avg. Value and 95% CI)	Diff (Days Separating τ from First Match)	b_1 (Angular Coefficient before τ, avg. Value and 95% CI)	b_2 (Angular Coefficient after τ, avg. Value and 95% CI)	a_1 (Intercept before τ, avg. Value and 95% CI)
Portugal	26.4 (2.4, 47.7)	8	0.03 (0.01, 0.05)	0.08 (0.08, 0.09)	6.28 (6.25, 6.32)
Russia	38.7 (38.4, 39.0)	24	0.03 (−0.03, −0.03)	0.00 (0.00, 0.00)	8.91 (8.90, 8.91)
Sweden	26.9 (7.9, 45.8)	10	−0.04 (−0.05, −0.02)	0.00 (−0.04, 0.05)	7.26 (7.15, 7.36)
Turkey	43.4 (43.1, 43.6)	29	−0.01 (−0.01, −0.01)	0.05 (0.05, 0.06)	8.93 (8.92, 8.94)
UK	52.6 (27.8, 28.6)	37	0.05 (0.05, −0.05)	−0.04 (−0.05, −0.04)	7.99 (7.98, 7.99)

For the sake of conciseness, we did not repeat, here again, the exercise to compute the halving/doubling times for those countries. Nonetheless, an interested reader could easily obtain those values by exploiting Formulas (2) and (3) in Section 2.2 and by using the correspondent data reported in Table 3.

3.2. Countries That Did Not Participate in the Tournament

While maintaining the pure observational nature of the inferences of our analysis about the effect of the tournament, we took advantage of another natural experiment, by observing what happened, during the tournament, in some 12 additional European countries that did not take part in the European football championship (considering the beginning of the tournament as the basis of our statistical observations).

This group was comprised of the following countries (with motivations for their choice reported in brackets): Greece and Ireland (great football traditions), Romania and Azerbaijan (hosting countries), Norway and Iceland (representatives of Northern Europe), Bulgaria and Moldova (representatives of Eastern Europe), Serbia and Bosnia (representatives of Balkans), and Latvia and Lithuania (largest countries representatives of Baltic Europe).

Needless to say, many other countries were left out. The motivations were manifold, ranging from their limited geographical dimensions (e.g., Malta, Faroe Islands, San Marino, Cyprus, Andorra, Montenegro, Kosovo, etc.) to geopolitical considerations, also in relationship with the game of football. For example: Georgia, Armenia, Kazakhstan, and Belarus are not famous for their international football traditions. Moreover, they are also well aligned with the contagion dynamics of one of their most influential neighboring countries, that is, Russia, which we had already examined.

The results of the application of our method to the above 12 countries are presented in Table 4. The 12 countries are listed based on the increasing value of *diff* (i.e., the number of days that separate the changepoint from the beginning of the tournament).

Here, it is important to remind what was already stated at the end of Section 2.2, that is: COVID-19 cases can show their biggest daily jumps 2–3 weeks after a particular mass event; hence, only those countries with inverting changes occurring in the time interval from 5–6 to 22–23 days after the beginning of the tournament were considered as those that have followed the pattern. This group is comprised of Greece, Azerbaijan, Ireland, Serbia—just 4 countries out of 12 (33%).

Table 4. Regression parameters for some 12 countries that did not participate in the tournament.

Country (Participating in the Tournament)	τ (Changepoint, avg. Value and 95% CI)	Diff (Days Separating τ from Beginning of Tournament)	b_1 (Angular Coefficient before τ, avg. Value and 95% CI)	b_2 (Angular Coefficient after τ, avg. Value and 95% CI)	a_1 (Intercept before τ, avg. Value and 95% CI)
Moldova	9.63 (6.52, 12.52)	−4	−0.06 (−0.10, −0.03)	0.01 (0.01, 0.02)	4.36 (4.21, 4.50)
Norway	16.56 (15.09, 17.98)	3	−0.05 (−0.06, −0.04)	−0.00 (−0.00, 0.00)	6.03 (5.98, 6.07)
Azerbaijan	24.16 (23.00, 25.32)	10	−0.08 (−0.08, −0.07)	0.06 (0.05, 0.06)	5.47 (5.41, 5.54)
Greece	26.03 (25.77, 26.31)	12	−0.06 (−0.06, −0.06)	0.07 (0.07, 0.07)	7.53 (7.51, 7.55)
Ireland	31.58 (30.53, 32.68)	18	−0.01 (−0.01, −0.01)	0.06 (0.05, 0.06)	6.05 (6.01, 6.08)
Serbia	34.84 (33.49, 36.20)	21	−0.05 (−0.05, −0.04)	0.05 (0.04, 0.06)	5.83 (5.79, 5.87)
Lithuania	39.12 (38.17, 40.08)	25	−0.08 (−0.08, −0.08)	0.09 (0.08, 0.10)	6.42 (6.39, 6.45)
Latvia	45.07 (41.54, 48.44)	31	−0.05 (−0.05, −0.05)	0.02 (−0.01, 0.05)	5.91 (5.87, 5.94)
Romania	37.60 (35.21, 39.90)	24	−0.06 (−0.06, −0.06)	0.04 (0.03, 0.05)	5.77 (5.72, 5.82)
Bosnia and Herzegovina	38.61 (36.16, 40.76)	25	−0.06 (−0.06, −0.05)	0.04 (0.02, 0.06)	4.63 (4.55, 4.70)
Bulgaria	39.82 (32.97, 44.20)	26	−0.04 (−0.04, −0.03)	0.04 (0.01, 0.06)	5.49 (5.44, 5.55)
Iceland	46.82 (44.85, 48.43)	33	−0.01 (−0.02, 0.00)	0.31 (0.27, 0.36)	1.40 (1.11, 1.70)

For all the other eight countries (67%), either their changepoint was premature (Norway and Moldova) or it came too late, precisely more than 23 days after the beginning of the tournament (Latvia, Lithuania, Romania, Bosnia, Bulgaria, and Iceland, in some cases, even without a clear case trend inversion, e.g., Bosnia).

As usual, with Figures 4 and 5, we portrayed a graphical representation of the same data of Table 4 for all the 12 countries of interest. Yet again, all the relevant information needed to interpret Figures 4 and 5 were inserted in the corresponding captions. We have used two separate figures, just for the sake of simplicity.

Finally, it is worth noting that we have not provided here again all the statistical information that we had computed for the countries participating in the tournament (e.g., various statistics, halving and doubling times, etc.). There is no precise motivation but that of brevity. Any interested reader could easily compute those statistics, with the data from Table 4. For example, halving and doubling times can be obtained by using the data from Table 4 along with the Formulas (2) and (3) of Section 2.2.

In conclusion, these final numbers have clearly shown that, while one could suppose that an increase in COVID-19 cases may have been an inevitable consequence of the general European situation in July 2021, the European football tournament, with its mass gatherings, played the important role of accelerator of this phenomenon, for many of its participating countries.

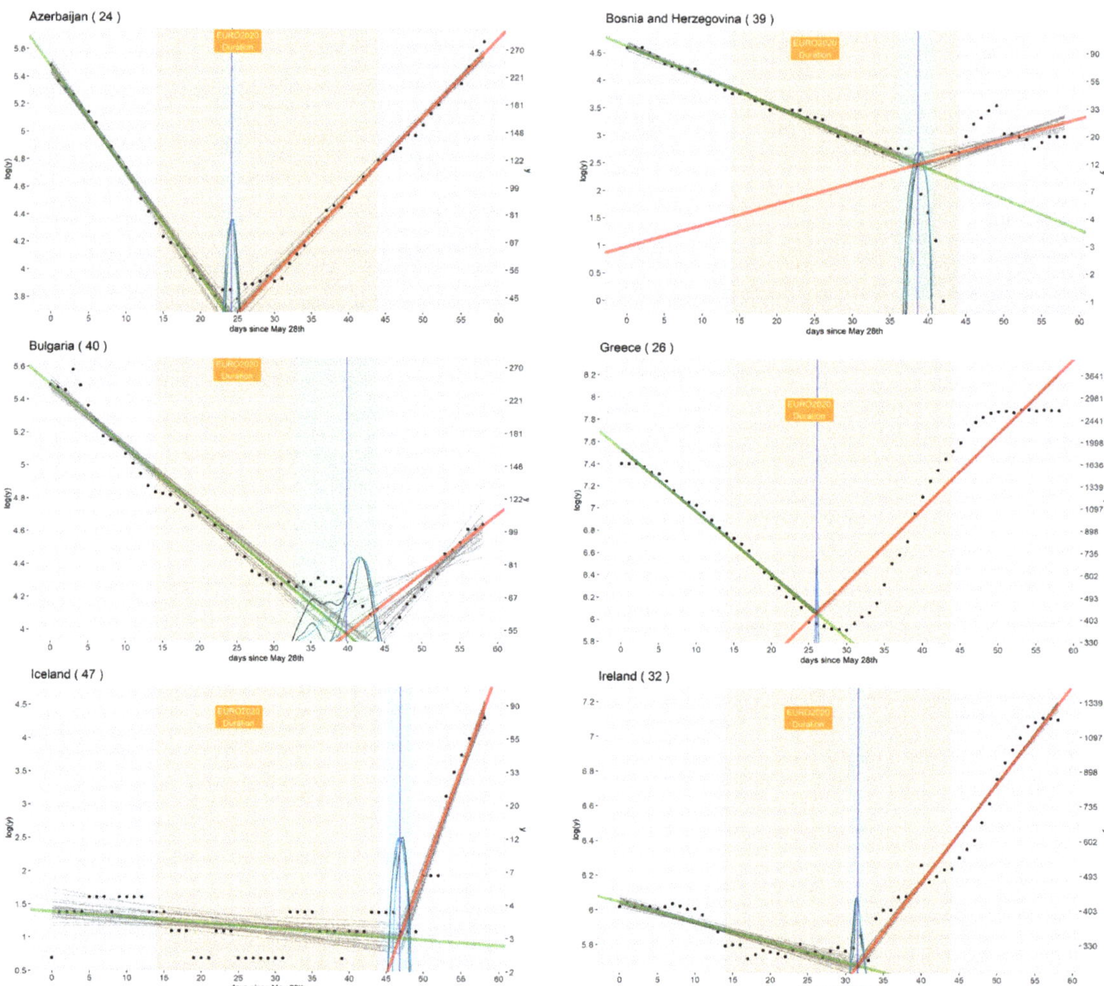

Figure 4. Azerbaijan, Bosnia, Bulgaria, Greece, Iceland, and Ireland not participating countries. Yellow space: duration of the tournament. Red vertical line: first match. Purple vertical line: last match. Green vertical line: last hosted match. Blue vertical line: changepoint. Blue space: CI amplitude for the changepoint. Green segment: case rate trend before the changepoint. Red segment: case rate trend after the changepoint. Blue bell-shaped peaks, grey segments, black dots, and rightmost and leftmost *y* axis: same as in previous figures.

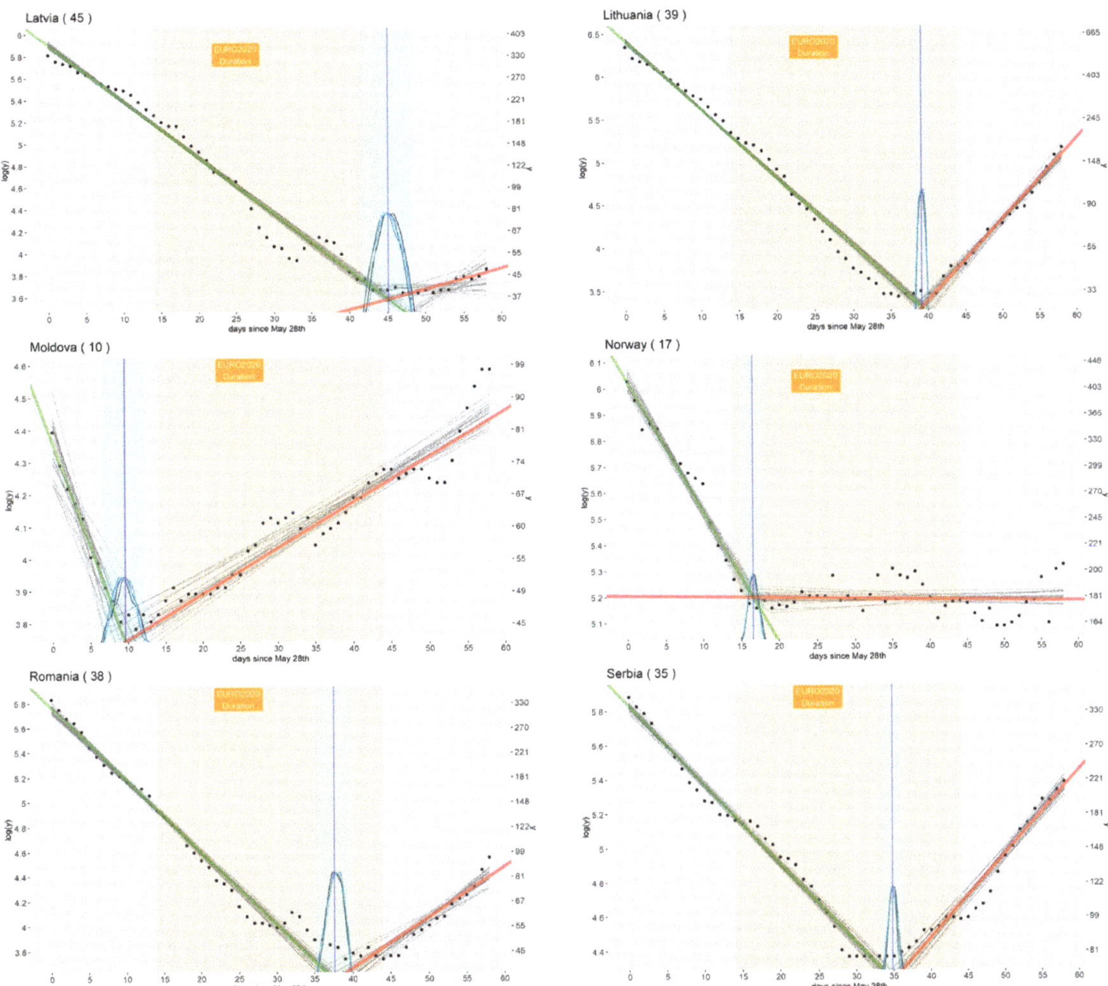

Figure 5. Latvia, Lithuania, Moldova, Norway, Romania, and Serbia not participating countries. Yellow space: duration of the tournament. Red vertical line: first match. Purple vertical line: last match. Green vertical line: last hosted match. Blue vertical line: changepoint. Blue space: CI amplitude for the changepoint. Green segment: case rate trend before the changepoint. Red segment: case rate trend after the changepoint. Blue bell-shaped peaks, grey segments, black dots, and rightmost and leftmost *y* axis: same as in previous figures.

4. Discussion and Conclusions

With this study, we found that, in 17 out of 22 (77%) countries involved in the 2020 European football championship, there has been a changepoint in the number of daily new SARS-COV-2 cases during the tournament, falling on average 14.97 days (95% CI 12.29–17.47) after the first match they played. Not only that, the case rate of the new daily infections was inverted for all these 17 countries, changing from a decreasing trend to an increasing one. We have quantified this inversion by measuring, for each national infection curve, the halving time before the change and the doubling time after it; they are respectively, on average: 18.07 days (95% CI 11.81–29.42) days and 29.10 days (95% CI 14.12–49.78).

There are five countries that break the pattern, and the presence of which could be seen as a first limitation of this study. Nonetheless, a careful consideration of the situation of these countries could provide a plausible explanation to this behavior. For example, it is evident that, for many of them (the UK, Russia, and Portugal), it is not possible to detect a changepoint in the infection rate which is coincidental with their participation in the tournament. This is because the inflation of the new COVID-19 cases was already in effect, in all these countries, when the tournament started, with the effect of the football championship probably absorbed into that inflation. The causes for this premature upturn of SARS-COV-2 cases are quite clear for the UK, which was the first European country to face the Delta variant. Portugal, instead, could have been the first European country to face the tourism impact, with many early tourists coming just from the UK. The situation in Russia, because of its enormous geographical extension, is, instead, too complex to look for a single explanation. Sweden, which reports COVID-19 numbers four days a week, entails a difficult interpretation, with our model not able to spot plausible changepoints. In regards of Sweden, it should not be forgotten that this was a country where very different strategies for managing the pandemic were adopted, without resorting, for example, to national lockdowns. Obviously, at the current stage of our research, no inference can be drawn regarding the existence of a relation between this fact and the results we achieved concerning this country. The situation for Turkey is different. It seems to follow the pattern, with an easily identifiable changepoint, coincidental with a reversion in the decrease/increase trend of the new COVID-19 cases. Nonetheless, this changepoint comes a bit too late (29 days after its first match). Hence, our decision was not to consider it as a further evidence in favor of the investigated link.

A second limitation of this study is that it ignored the possible effects of other confounding factors that could have played a role. Unfortunately, there are too many, and they are also too country specific, in many cases, to be considered as a whole. Nonetheless, the following two facts should also be considered. A general trend toward the decrease in the new daily SARS-COV-2 cases had already begun during the beginning of the 2021 spring, in almost all the considered countries, as an effect of the vaccination. In response to the benefits of the vaccines, almost all these European countries had consequently begun to lift the restrictions that were imposed to combat the third wave of the contagion. This happened well before the beginning of the tournament, and without any evident effect in terms of an upturn of new SARS-COV-2 cases (with the only exception of the already-discussed situation in the UK). It is a matter of fact, instead, that many infection clusters have surged in Europe during the football tournament. At the end, despite many possible country-specific confounding factors that could have played a role, our study has revealed that the temporal coincidence between the tournament and the inverting trend of the infections in many participating countries is an issue that cannot go unnoticed.

A third limitation touches more upon the mathematical and statistical nature of our analysis. We have already anticipated that our study is purely observational, without any possibility to demonstrate the existence of a clear relationship of cause and effect. Our intent was simply that of enquiring if all the mass gatherings following the football matches could have correlated with the virus resurgence in many European countries. For this reason, to study the plausibility of the correlation of interest, we have developed a simple model (similar to that employed in [20]) that does not possess the ambition of being exhaustive in the representation of the COVID-19 dynamics [21]; instead, it is very effective in detecting a changepoint in the infection curves, with the two corresponding slopes (before and after it) with which the decrease/increase case trends can be analyzed.

Finally, with an additional experiment, we have also demonstrated that the number of countries that follow the pattern falls down from 77% to 33% if we consider European countries that did not take part in the tournament. At the end, we can conclude that the results of our analysis are compatible with the hypothesis that most of the countries involved in the European football championship have seen a rise in the number of new SARS-COV-2 cases, or a slowdown in the fall, temporally coincident with their participation.

While this study has no ability to establish a final causal relationship, we think that the tournament, with its mass gatherings inside and outside the stadiums, has surely had an acceleration effect, that, coupled with the release of restrictions, could have given a contribution to ignite a new wave of the COVID-19 spread.

Author Contributions: Conceptualization, L.C. and M.R.; methodology, L.C. and M.R.; software, L.C.; validation, L.C. and M.R.; formal analysis, L.C.; investigation, M.R.; resources, M.R.; data curation, L.C.; writing—original draft preparation, L.C. and M.R.; writing—review and editing, L.C. and M.R.; visualization, L.C.; supervision, M.R.; project administration, M.R.; funding acquisition, M.R. Both authors have read and agreed to the published version of the manuscript.

Funding: This research received no external funding.

Data Availability Statement: All data regarding COVID-19 infections used in this paper are available on the online repository termed Our World in Data, accessible at the following URL: https://github.com/owid/covid-19-data/blob/master/public/data/README.md (accessed on 16 August 2021).

Conflicts of Interest: The authors declare no conflict of interest.

References

1. Thomasson, E. German Minister Chides 'Irresponsible' UEFA over Euro 2020 Crowds. Reuters. 2021. Available online: https://www.reuters.com/world/europe/german-minister-slams-uefas-decision-fuller-stadiums-2021-07-01/ (accessed on 28 July 2021).
2. World Health Organization. Statement by Dr Hans Henri P. Kluge, WHO Regional Director for Europe. 2021. Available online: https://www.euro.who.int/en/media-centre/sections/statements/2021/statement-covid-19-the-stakes-are-still-high (accessed on 28 July 2021).
3. Henley, J.; Rankin, J. COVID: Euro 2020 Crowds 'a Recipe for Disaster', Warns EU Committee", The Guardian. 2021. Available online: https://www.theguardian.com/world/2021/jul/01/covid-euro-2020-crowds-a-recipe-for-disaster-warns-german-minister0 (accessed on 28 July 2021).
4. UEFA. Euro 2020 Key Information for Spectators. 2021. Available online: https://www.uefa.com/uefaeuro-2020/news/025b-0ef33753d7d0-100629325be2-1000--key-information-for-euro-spectators/ (accessed on 28 July 2021).
5. Italian Associated Press Agency (ANSA). Cluster of 91 COVID-19 Cases Linked to Euro 2020 Game. 2021. Available online: https://www.ansa.it/english/news/general_news/2021/07/16/cluster-of-91-covid-19-cases-linked-to-euro-2020-game_84349124-e130-453b-ade2-8b7136bd8993.html (accessed on 28 July 2021).
6. Kington, T. Italy's Euro 2020 Victory Tour Sent Rome Cases Rocketing. *The Times*, 22 July 2021, pp. 1–4. Available online: https://www.thetimes.co.uk/article/italys-euro-2020-victory-tour-sent-rome-cases-rocketing-r6m667r0b (accessed on 28 July 2021).
7. Peltier, E. Crowds for European Championship Soccer Games Are Driving Infections, the W.H.O. Says. *New York Times*, 1 July 2021, pp. 1–2. Available online: https://www.nytimes.com/2021/07/01/world/europe/euro-2020-covid-outbreak.html (accessed on 28 July 2021).
8. Skydsgaard, N.; Gronholt-Pedersen, J. Euro 2020 Crowds Driving Rise in COVID-19 Infections, Says WHO. Reuters. 2021. Available online: https://www.reuters.com/world/europe/who-warns-third-coronavirus-wave-europe-2021-07-01/ (accessed on 28 July 2021).
9. Schumacher, Y.O.; Tabben, M.; Hassoun, K.; Al Marwani, A.; Al Hussein, I.; Coyle, P.; Abbassi, A.K.; Ballan, H.T.; Al-Kuwari, A.; Chamari, K.; et al. Resuming professional football (soccer) during the COVID-19 pandemic in a country with high infection rates: A prospective cohort study. *Br. J. Sports Med.* **2021**, 1–11. [CrossRef]
10. Kochańczyk, M.; Grabowski, F.; Lipniacki, T. Super-spreading events initiated the exponential growth phase of COVID-19 with $\mathcal{R}0$ higher than initially estimated. *R. Soc. Open Sci.* **2020**, *7*, 200786. [CrossRef] [PubMed]
11. Weed, M.; Foad, A. Rapid Scoping Review of Evidence of Outdoor Transmission of COVID-19. *medRxiv* **2020**, 1–16. Available online: https://www.medrxiv.org/content/10.1101/2020.09.04.20188417v2 (accessed on 16 August 2021).
12. Cereda, D.; Tirani, M.; Rovida, F.; Demicheli, V.; Ajelli, M.; Poletti, P.; Trentini, F.; Guzzetta, G.; Marziano, V.; Barone, A.; et al. The early phase of the COVID-19 outbreak in Lombardy, Italy. *arXiv* **2020**, arXiv:2003.0932.
13. Mercker, M.; Betzin, U.; Wilken, D. What influences COVID-19 infection rates: A statistical approach to identify promising factors applied to infection data from Germany. *medRxiv* **2020**, 1–12. Available online: https://www.medrxiv.org/content/10.1101/2020.04.14.20064501v1 (accessed on 16 August 2021).
14. Signorelli, C.; Odone, A.; Riccò, M.; Bellini, L.; Croci, R.; Oradini-Alacreu, A.; Fiacchini, D.; Burioni, R. Major sports events and the transmission of SARS-CoV-2: Analysis of seven case-studies in Europe. *Acta Biomed.* **2020**, *91*, 242–244. [CrossRef]
15. El Hassan, N. Deloitte's Sports Business Group Estimates That Football Money League Clubs Will Miss out on Revenue of over €2 Billion by End of the 2020/21 Season Due to the COVID-19 Pandemic. 2021. Available online: https://www2.deloitte.com/xe/en/pages/about-deloitte/articles/deloittes-sports-business-group-estimates-football-money-league-miss-out-revenue-over-2euros-billion-end-202021-due-covid.html (accessed on 28 July 2021).

16. Our World in Data. COVID-19 GitHub Repository. 2021. Available online: https://github.com/owid/covid-19-data/blob/master/public/data/README.md (accessed on 28 July 2021).
17. Wikipedia. European Football Championship 2020. Available online: https://en.wikipedia.org/wiki/UEFA_Euro_2020 (accessed on 28 July 2021).
18. Lindeløv, J.K. mcp: An R Package for Regression with Multiple Change Points. 2020. Available online: https://osf.io/fzqxv/ (accessed on 28 July 2021).
19. Sebastiani, G.; Palù, G. COVID-19 and School Activities in Italy. *Viruses* **2020**, *12*, 1339. [CrossRef] [PubMed]
20. Casini, L.; Roccetti, M. Reopening Italy's schools in September 2020: A Bayesian estimation of the change in the growth rate of new SARS-CoV-2 cases. *BMJ Open* **2021**, *11*, e051458. [CrossRef]
21. Tosi, D.; Campi, A. How Data Analytics and Big Data Can Help Scientists in Managing COVID-19 Diffusion: Modeling Study to Predict the COVID-19 Diffusion in Italy and the Lombardy Region. *J. Med. Internet Res.* **2020**, *22*, e21081. [CrossRef] [PubMed]

future internet

MDPI

Article

How Schools Affected the COVID-19 Pandemic in Italy: Data Analysis for Lombardy Region, Campania Region, and Emilia Region

Davide Tosi [1,*] and Alessandro Siro Campi [2]

1   Department of Theoretical and Applied Sciences (DiSTA), University of Insubria, 21100 Varese, Italy
2   Department of Electronics, Information and Bioengineering (DEIB), Politecnico Milano, 20133 Milano, Italy; alessandro.campi@polimi.it
*   Correspondence: davide.tosi@uninsubria.it

Abstract: Background: Coronavirus Disease 2019 (COVID-19) is the main discussed topic worldwide in 2020 and at the beginning of the Italian epidemic, scientists tried to understand the virus diffusion and the epidemic curve of positive cases with controversial findings and numbers. Objectives: In this paper, a data analytics study on the diffusion of COVID-19 in Lombardy Region and Campania Region is developed in order to identify the driver that sparked the second wave in Italy. Methods: Starting from all the available official data collected about the diffusion of COVID-19, we analyzed Google mobility data, school data and infection data for two big regions in Italy: Lombardy Region and Campania Region, which adopted two different approaches in opening and closing schools. To reinforce our findings, we also extended the analysis to the Emilia Romagna Region. Results: The paper shows how different policies adopted in school opening/closing may have had an impact on the COVID-19 spread, while other factors related to citizen mobility did not affect the second Italian wave. Conclusions: The paper shows that a clear correlation exists between the school contagion and the subsequent temporal overall contagion in a geographical area. Moreover, it is clear that highly populated provinces have the greatest spread of the virus.

Keywords: COVID-19; SARS-CoV-2; data analytics; schools' impact; Google mobility impact

Citation: Tosi, D.; Campi, A.S. How Schools Affected the COVID-19 Pandemic in Italy: Data Analysis for Lombardy Region, Campania Region, and Emilia Region. *Future Internet* **2021**, *13*, 109. https://doi.org/10.3390/fi13050109

Academic Editor: Michael Sheng

Received: 24 March 2021
Accepted: 26 April 2021
Published: 27 April 2021

Publisher's Note: MDPI stays neutral with regard to jurisdictional claims in published maps and institutional affiliations.

1. Introduction

Data analysis [1–3] has proved to be of fundamental importance for studying and predicting the behavior of the pandemic of SARS-CoV2 and COVID-19, in order to intervene promptly and stem its spread [4–6]. The school opening has been a hotly debated topic nationwide and worldwide [7–9], with at one side scientists that consider schools safe and on the other side scientists who consider schools unsafe and unsecured. In our opinion, school is not a safe environment by definition, but it must be made safe taking serious actions with rigorous protocols and structural interventions as described in [10,11]. Effects of schools opening and the propagation of COVID-19 are described in other countries, such as in [12] where the effects of school openings on hospitalization in USA are modeled, or in [11] where the authors explain how UK schools are causing COVID-19 spreading and how to act to reduce their impact.

The data (shown in Table 1) on the growth of infections by age groups from the beginning of September to March that are published weekly in the epidemiological reports of the ISS (Istituto Superiore Sanità) [www.iss.it] (accessed on 27 April 2021), indicate that the age group 0–9 have had a growth between 6 and 10 times higher than all other ages. (Please note that following the ISS indication, under <19 young population are mostly asymptomatic with a percentage of 75%. Hence, the ratio of these cases in the younger population is probably much higher than in the elder population.)

Table 1. The trend of infections between 29 December 2020 and 10 March 2021.

Age	Number of Cases 12/29/2020	Number of Cases 3/10/2021	Percentage Growth
0–9	78,664	144,301	83.44
10–19	170,048	277,785	63.36
20–29	245,458	367,308	49.64
30–39	251,226	382,754	52.35
40–49	326,571	494,423	51.40
50–59	368,635	545,225	47.90
60–69	229,200	344,498	50.30
70–79	172,071	255,511	48.49
80–89	149,953	209,503	39.71

The data show that from 29 December 2020 to 10 March 2021, the infections increased by 83.44% in the age group between 0 and 9 years. Additionally, 63.55% in the 10–19 age range. The school age is therefore the one where the contagion has grown a little more. The third group for growth is the one between 30 and 39 years, with 52.35% and almost all the other age groups are below 50% growth. The older age groups registered the lowest percentage growth: between the ages of 80 and 89, the contagion grew by 39.71% over the period, and over 90 years even less 31.28%. These data disprove the idea that the problem was that of transport, which essentially concerned the high schools (the youngest ones mostly go to school on foot or are accompanied by their parents by private means).

Looking at the data and statements of neighboring countries and with demographic characteristics similar to ours, the situation is already well defined on how much schools are drivers of contagion:

- In France schools and universities have been indicated as the first factor in active outbreaks [source: Sante Publique France];
- In the UK, primary and secondary school, after careful tracing, was in third place as number of reports [source: NHS Test and Trace UK];
- In Germany, the school was recently declared to be at high risk in some statements made by A. Merkel herself in early February 2021;
- Several papers (recently appeared in the Lancet, Nature and Science), albeit with all the stated limits in the works, show that the closure of schools is the second most impacting factor, as NPI (Non-Pharmaceutical Interventions), on the reduction of the contagiousness index Rt [13,14];
- Further preliminary analyses have been carried out on the Piedmonts Region and national territory by researcher A. Ferretti and a clear relation on the increase of cases and the school opening is reported [15];
- The Lazio Region was driven to an emergency state due to school contagion impact [16];
- During the 19 March 2021 Press Conference, the Belgium Prime Minister said: "From contact analyses, we can also see that schools are key places where many infections happen," De Croo said. "Children are infected there, take the virus home, possibly infect their parents, who may infect their colleagues if they are still going to work, and so the chain continues" [17];
- We also remind you that 75% of the positives in the youth age group under 19 are asymptomatic, therefore, are unaware carriers of the virus within family walls [source ISS];
- In the week of mid-February 2021 alone, we collected more than 50 newspaper articles (headings national and local) that highlight outbreaks in Italian schools [18];

- Recent statements by the ISS (Higher Institute of Health) Director G. Rezza, dated 26 February 2021, highlight the problem of numerous outbreaks in Italian schools.

We believe the distinction between the school environment per se or extended to include the public transport and the dynamics of entering/leaving the school has no meaning. At the moment, the main contribution of the school to viral circulation must be analyzed and quantitatively assessed.

Obviously, it remains of fundamental importance to determine what risks are exposed to children with school closures, which certainly impacts on mental health, cognitive development and which are fundamental in developmental age and, consequently, arrive at risk-weighted decisions, as described in [19–21] where the authors highlight the psychological impact school closures may have on young people.

At the end of August, we presented a predictive model to show how the second wave in Italy was practically already started. The model estimated a relative peak around the 7/8 of September and then a slight decline in slowdown waiting to see the strong impact, within two weeks, coming from the schools reopening (on 14 September). If there now we look back (see Figure 1), we clearly observe that the exponential explosion of the contagion in Italy started exactly on 28 September, so exactly two weeks after the reopening of the Italian schools.

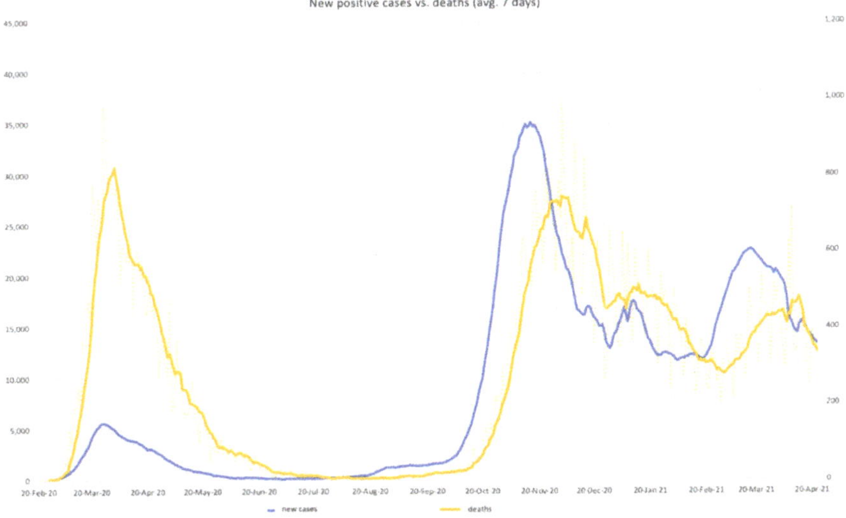

Figure 1. Observed real curves for new daily cases and deaths in Italy.

Our model was based on the hypothesis that schools are an important driver of contagion. Furthermore, the major impact is to be considered in the contagion that then happens at a second layer inside the family context, leading after about two incubation cycles of the virus. This explains why we used a time lag of 14 days, in our predictive model. Even if it is not our aim to "blame" children or teachers for these infections and we watched the school operators doing their utmost in the summer to find solutions to secure the school environment as much as possible, we cannot be blind and avoid seeing that the virus finds fertile ground for contagion in closed environments, very populated, poorly ventilated, as are our school environments which are not among the most modern in Europe. Therefore, to think that the school is a safe environment, by definition, is wrong because it has caused and will be the cause of uncontrolled virus spread.

In this paper, we want to analyze the few official MIUR (Ministry of Instruction, University and Research) data available on contagion at schools, to understand whether and how much the school may have impacted on the territorial contagion.

2. Materials and Methods

In December, the MIUR published an official dataset relating to infections in schools (joining different data collection) for the period 14 September 2020–30 October 2020 [22]. The report spoke of approximately 65,000 Italian positive cases identified in the time window 14 September–30 October (we are only talking about primary and lower secondary school, because most of the high schools were in any case remote). The available data count 65,000 cases for the whole Italy, but they are underestimated because not all Italian schools have participated in this tracking activity, and not all schools have released their data to the ministry. It should also be considered that 75% of those under 19 years old are asymptomatic [www.iss.it] (accessed on 27 April 2021), and this large slice of young people is lost in the tracing activity. In total, 65,000 cases out of 360,000 total cases [23,24] detected in the same period is a considerable percentage of 18% of the total. Furthermore, the major impact is to be considered in the contagion that then arose in the second instance within the family walls, leading after about two incubation cycles of the virus, to an uncontrolled growth of the curves (the one that we observed from 28 September 2020 onwards in our predictive model and in Figure 1).

In order to understand the relation between schools and global infection, we considered the data officially released by the MIUR and we carried out a correlation analysis on the Lombardy Region (RL) and Campania Region (RC), two regions that have adopted two different policies of opening and closing schools. RL is characterized by 12 provinces for a total of 10 M inhabitants, while RC has 5 provinces with 5.8 M inhabitants. Moreover, we extended this analysis to a third Italian region: Emilia Romagna Region (REm) that is characterized by 9 provinces and a total population of 4.5 M inhabitants. At the end, the study covered 24 provinces out of 107 Italian provinces, and 20.3 M inhabitants out of 60.3 M total Italian inhabitants.

The three regions applied the following opening/closure strategies:

- RL reopened all primary and secondary schools in presence at 14 September 2020 (high level secondary school with 50% attendance and 50% online) [25];
- RC reopened all primary and secondary schools in presence at 24 September 2020 (high level secondary school with 50% attendance and 50% online) and then all levels were closed in advance starting from the October 16 and until 13 November [26];
- REm reopened all primary and secondary schools in presence at 14 September 2020 (high level secondary school with 50% attendance and 50% online).

Specifically, a twofold correlation study was conducted:

1. between school contagion index (both total and separate for primary and secondary school, respectively) and an index of global contagion at the provincial level (both for RL and RC). The correlation study was done with a global contagion index on the reference period from 14 September 2020 to 30 October 2020 and also considering the first two weeks after the reopening of schools (from 14 September to 28 September, where the contagion theoretically should not be detectable, given the latency time between positivity and the onset of symptoms and related diagnostic screening) then in the following two weeks (from 28 September to 12 October, when it is likely that contagion was triggered in schools and then it potentially spreads in the intrafamily context), and after four weeks of spreading;
2. between contagion index and mobility indexes derived from the COVID-19 Google Community Mobility Report [27], where mobility data at regional and national level in different sectors (e.g., mobility near parks and public gardens, pharmacies, at work level, train stations, residential, etc.) were analyzed.

We computed the correlation index by using the Pearson correlation index (CI), as:

$$\text{correl\_index}\,(x, y) = \frac{\sum (x - \bar{x})(y - \bar{y})}{\sqrt{\sum (x - \bar{x})^2 \sum (y - \bar{y})^2}} \tag{1}$$

F-Test was then conducted on the dataset to determine if there is a significant difference between the means of two groups and to understand the statistical significance of our findings. Linear regression models and the R^2 coefficient of determination were also discussed.

Summarizing, all the datasets used in this study are:

- Official Positive Cases at Italian Schools released by MIUR Ministry [22];
- Official Positive Cases in Italy by Department of Civil Protection [24];
- Rt Dataset by University of Insubria [23];
- Mobility Data by Google Community [27].

3. Results

3.1. Comparing Lombardy, Campania and Emilia Romagna Contagion Indexes

As for the Lombardy Region (RL), the identified cases are 13,967 out of 88,412 total cases (15.8%), in the reference period September 14th–October 30th. In the Campania Region (RC), the cases identified are 4620 about 42,815 total cases (10.8%). It is important to recall that the Lombardy Region and the Campania Region have used in October different school policies, the first leaving primary and secondary schools open in attendance and high secondary schools at 50% in attendance [25], while RC intervening instead with targeted closures: schools opened at September 24 (secondary schools at 50% in attendance), and then all levels were closed in advance starting from the October 16 and until November 13 [26].

Let us now focus on the data of the Lombardy Region at the provincial level (see Figure 2 for numerical details for all the RL provinces: VA Varese, SO Sondrio, PV Pavia, MN Mantova, MI Milan, MB Monza and Brianza, LO Lodi, LC Lecco, CR Cremona, CO Como, BS Brescia, and BG Bergamo). If we calculate a "school contagion" index and a "global contagion" index (normalized on the ISTAT2020 population [http://demo.istat.it] (accessed on 27 April 2021), for each province as: cases/1000 inhabitants), it is interesting to note that there is a strong correlation between a high rate of contagion in school and high contagion rate then at the provincial level. This correlation is not found instead by looking at, for example, the population density as another variable. As an example, let us take the case of Varese (VA): VA was one of the Lombardy provinces that was mostly impacted by the second COVID-19 wave. VA has the highest school contagion rate (together with MB) and a very high global contagion rate, as depicted in Figure 2.

Let us now consider what happened in the first two weeks after the school reopening (14 September–28 September) where theoretically the effects of the school were just beginning to be visible: it is noted how there is no correlation between the school contagion index and the global contagion index at two weeks (CI = −0.10). If, on the other hand, we look at the correlation between school contagion and the index of contagion two weeks after the reopening of schools, we have a clear correlation with CI = 0.69, which rises further considering the effect after four weeks with CI = 0.80. For the entire reference period the CI rises to CI = 0.89) as clearly depicted in the scatter plot of Figure 3 (the linear regression trend line as the R^2 coefficient of determination with a very high value $R^2 = 0.94$). The correlation indices are identical if the school contagion data is separated between primary and secondary school. It is interesting to observe the correlation between the contagion index and the population density (CI = 0.60) but not between the school index and the population density (CI = 0.37). If, on the other hand, we observe the correlation between the contagion index and the mobility indexes, we note that the lower mobility registered with government restrictions and DPCM (Decreto del Presidente del consiglio dei ministri) does not have an interdependence relation with the contagion more or less accentuated in the various provinces of RL, with the exception of transit station mobility that shows a correlation CI = −0.57.

	School index	Primary School index (cases/1000 pop.)	Secondary School index	Contagion index (sept. 14 to oct. 30)	Contagion index first 2 weeks (sept. 14 to sept. 28)	Contagion index after 2 weeks (sept. 28 to oct. 12)	Contagion index after 4 weeks (oct. 12 to oct. 26)
VA	2.6	1.7	0.9	11.2	0.24	0.69	4.57
SO	1.2	0.8	0.4	6.5	0.14	0.55	2.62
PV	0.7	0.5	0.2	7.7	0.41	0.76	3.66
MN	0.7	0.5	0.2	4.1	0.16	0.37	1.84
MI	1.7	1.0	0.6	12.8	0.34	1.04	6.64
MB	2.7	2.0	0.8	13.1	0.08	0.96	5.82
LO	1.1	0.8	0.3	6.9	0.20	0.43	3.69
LC	1.2	0.7	0.5	6.5	0.13	0.44	3.04
CR	0.7	0.4	0.3	4.7	0.08	0.37	2.18
CO	1.5	1.1	0.4	8.6	0.12	0.45	3.74
BS	0.4	0.3	0.1	3.5	0.23	0.39	1.58
BG	0.4	0.3	0.1	2.5	0.15	0.32	1.02

Parks Mobility	Retail - Recreation Mobility (sept. 1 to oct. 15 - avg value)	Transit Station Mobility	Population ISTAT 2020	Pop. Density (pop./kmq)
20.38	-10.48	-35.80	892532	744.95
19.33	-15.98	-21.51	180941	56.62
14.66	-7.00	-27.07	546515	184.1
32.81	-10.96	-12.87	411062	175.56
19.39	-20.76	-32.09	3279944	2081.64
33.41	-11.52	-23.50	878267	405.41
-2.25	-1.54	-20.07	230607	294.52
38.42	-5.37	-8.17	337087	418.42
19.28	-10.70	-16.35	358347	202.4
76.89	-9.96	-6.78	603828	472.1
80.13	-8.67	-6.85	1268455	265.06
33.67	-12.04	-22.30	1116384	405.24

LOMBARDY REGION
(school opening sept. 14)

Corr. Index	❌	0.89	school index vs. contagion index
Corr. Index	❌	0.88	primary school index vs. contagion index
Corr. Index	❌	0.87	secondary school index vs. contagion index
Corr. Index	✅	-0.10	school index vs. contagion index first 2 weeks
Corr. Index	❌	0.69	school index vs. contagion index after 2 weeks
Corr. Index	❌	0.80	school index vs. contagion index after 4 weeks
Corr. Index	❌	0.60	pop. density vs. contagion index
Corr. Index	✅	-0.21	contagion index vs. Parks mobility
Corr. Index	✅	-0.32	contagion index vs. Retail mobility
Corr. Index	✅	-0.57	contagion index vs. Transit station mobility
Corr. Index	🟠	0.37	school index vs. pop. density

Figure 2. Lombardy Region data set for school contagion index and overall contagion. Correlation study.

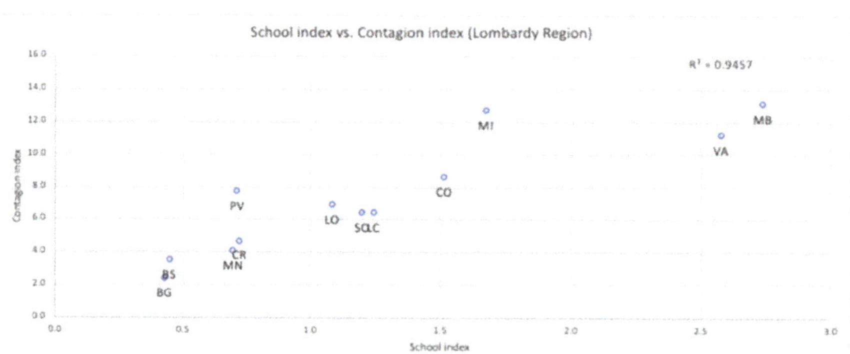

Figure 3. Lombardy Region scatter plot for school contagion index vs. overall contagion index.

Since the number of datapoints is limited, we tested the statistical significance with F-Test (α = 0.05) and we obtained the following values: F = 21.20 with a P(F) = 7.94×10^{-6} and an F critical = 2.82 (gdl = 11), so the null hypothesis is rejected (since F calculated is greater than the F critical).

As for Campania Region (Figure 4) and its five provinces (SA Salerno, NA Napoli, CE Caserta, BN Benevento, AV Avellino), schools reopened two weeks later than RL and were closed in advance, starting from 16 October and until 13 November. We observed the following statistical behavior between the contagion variables in the schools and the subsequent contagion in the regional provinces: in the first few days after the schools re-opening (28 September to 12 October) the correlation index is equal to CI = 0.51 and then increased after two more additional weeks to the strong value CI = 0.93 (i.e., showing the clear impact of schools reopening). The global correlation index is equal to CI = 0.47, with a behavior that is in line with the one observed for Lombardy Region. There is only a temporal shift ahead, since RC reopened schools two weeks later than RL. Moreover, also for RC, a strong correlation exists between the contagion index and the population density (CI = 0.74). As for mobility data, a strong correlation index is detected only between the contagion index and the retail and recreation mobility (CI = 0.86), thus suggesting that this mobility factor may have had an impact to the spread of COVID-19, too.

School index	Primary School index (cases/1000 pop.)	Secondary School index	Contagion index (sept. 14 to oct. 30)	Contagion index first 2 weeks (sept. 14 to sept. 28)	Contagion index after 2 weeks (sept. 28 to oct. 12)	Contagion index after 4 weeks (oct. 12 to oct. 26)
SA	0.5	0.3	0.2	3.7	0.21	0.49	2.00
NA	1.0	0.6	0.4	9.2	0.65	1.64	4.44
CE	0.8	0.6	0.3	8.1	0.54	1.06	3.64
BN	0.8	0.4	0.4	2.6	0.36	0.58	3.15
AV	0.2	0.2	0.1	4.9	0.27	0.86	1.89

Parks Mobility	Retail - Recreation Mobility (sept. 1 to oct. 15 - avg value)	Transit Station Mobility	Population ISTAT 2020	Pop. Density (pop./kmq)
72.33	-2.4	6.27	1092779	221
17.15	-11.06	-22.88	3082905	2615
15.45	-7.17	-9.58	922171	348
14.14	-8.21	-23.63	274080	132
7.8	-8.45	-15.39	413926	148

	Corr. Index	0.47	school index vs. contagion index
	Corr. Index	0.61	primary school index vs. contagion index
CAMPANIA REGION	Corr. Index	0.23	secondary school index vs. contagion index
(school opening sept. 24)	Corr. Index	0.81	school index vs. contagion index first 2 weeks (school not yet reopened)
	Corr. Index	0.51	school index vs. contagion index after 2 weeks (4 days after school opening)
	Corr. Index	0.93	school index vs. contagion index after 4 weeks (2 weeks after school opening)
	Corr. Index	0.74	pop. density vs. contagion index
	Corr. Index	0.59	school index vs. pop. density
	Corr. Index	-0.32	contagion index vs. Parks mobility
	Corr. Index	0.86	contagion index vs. Retail mobility
	Corr. Index	-0.21	contagion index vs. Transit station mobility

Figure 4. Campania Region data set for school contagion index and overall contagion. Correlation study.

To reinforce our findings, we extended our analysis evaluating also the Emilia Romagna Region and its nine provinces (BO—Bologna, FC—Forlì-Cesena, FE—Ferrara, MO—Modena, PC—Piacenza, PR—Parma, RA—Ravenna, RE—Reggio Emilia, RN—Rimini). For this region, the identified cases are 3050 out of 19,670 total cases (15.5%), in the reference period 14 September–30 October, with a similar % to RL. Moreover, REm followed an opening/closure strategy such as the one applied to RL. As depicted in Figures 5 and 6, correlation between the school contagion index and the overall contagion index per each province increases over time, starting from an inverse correlation at the beginning of school reopening CI = −0.50 to an index of CI = 0.41 after two weeks, and CI = 0.69 after four weeks since school reopening. The index for the overall period is CI = 0.76 with the secondary school impacting more than the primary school (CI = 0.86 vs. CI = 0.58). Hence, also in this case, the detected behavior is consistent with the one observed in RL and RC.

Since the number of datapoints is limited, we tested the statistical significance with F-Test ($\alpha = 0.05$) and we obtained the following values: F = 23.16 with a P(F) = 9.27×10^{-5} and an F critical = 3.43 (gdl = 8), so the null hypothesis is rejected (since F calculated is greater than the F critical).

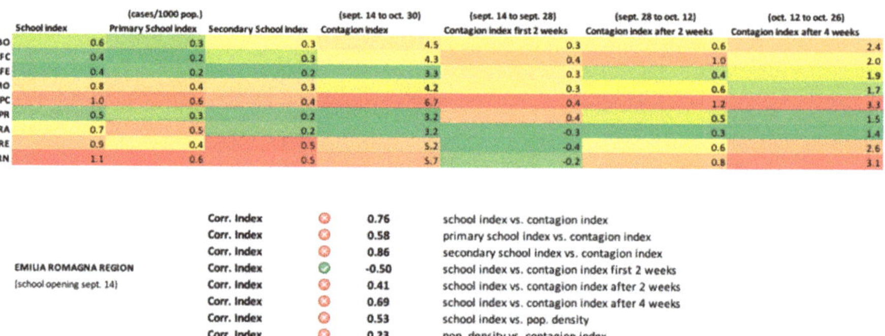

| | School index | (cases/1000 pop.) | | (sept. 14 to oct. 30) | (sept. 14 to sept. 28) | (sept. 28 to oct. 12) | (oct. 12 to oct. 26) |
		Primary School index	Secondary School index	Contagion index	Contagion index first 2 weeks	Contagion index after 2 weeks	Contagion index after 4 weeks
BO	0.6	0.3	0.3	4.5	0.3	0.6	2.4
FC	0.4	0.2	0.3	4.3	0.4	1.0	2.0
FE	0.4	0.2	0.2	3.3	0.3	0.4	1.9
MO	0.8	0.4	0.3	4.2	0.3	0.6	1.7
PC	1.0	0.6	0.4	6.7	0.4	1.2	3.3
PR	0.5	0.3	0.2	3.2	0.4	0.5	1.5
RA	0.7	0.5	0.2	3.2	-0.3	0.3	1.4
RE	0.9	0.4	0.5	5.2	-0.4	0.6	2.6
RN	1.1	0.6	0.5	5.7	-0.2	0.8	3.1

	Corr. Index	0.76	school index vs. contagion index
	Corr. Index	0.58	primary school index vs. contagion index
	Corr. Index	0.86	secondary school index vs. contagion index
EMILIA ROMAGNA REGION	Corr. Index	-0.50	school index vs. contagion index first 2 weeks
(school opening sept. 14)	Corr. Index	0.41	school index vs. contagion index after 2 weeks
	Corr. Index	0.69	school index vs. contagion index after 4 weeks
	Corr. Index	0.53	school index vs. pop. density
	Corr. Index	0.23	pop. density vs. contagion index

Figure 5. Emilia Romagna Region data set for school contagion index and overall contagion. Correlation study.

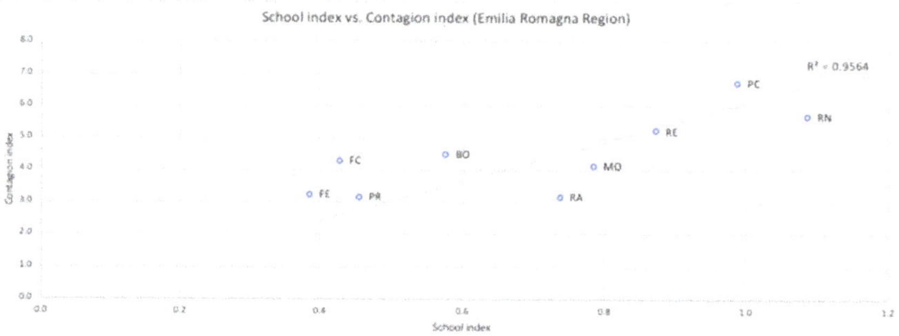

Figure 6. Emilia Romagna scatter plot for school contagion index vs. overall contagion index.

3.2. Reproduction Number (Rt) and Contagion Curves Evaluation

If we take also a look at the contagion curve (new daily positive cases), see Figure 7, RC, which was the region that applied the more restrictive policies for schools, was able to invert the trend of new daily positive cases earlier than the other two regions. Moreover, it is interesting to observe that also the ascent trend was less steep than for RL and REm, with the average doubling time (in the number of positive cases) equal to 8 days (3.4 days for RL and 6 days for REm).

Moreover, we computed the Rt (reproduction number) for all the Italian regions [23]. The Rt estimation was conducted by using the Time-Dependent method by Wallinga and Teunis [28] with a time aggregation level equal to 10 days, to understand the impact of schools reopening on the Rt trend. All regional and provincial trends are reported in our web site: www.covid19-italy.it (accessed on 27 April 2021). The trend is depicted in Figure 8: it is clear that RC was able to contain the reproduction number to a low peak value Rt = 1.1 while REm and RL have higher peak values of Rt = 1.4 and Rt = 1.5, respectively. Moreover, RC was the first one to reach the guard value Rt = 1.0 (9 November2020) probably due to the prompt school closures. In Table 2, we also summarize for all the Italian regions (categorized by the date of school opening) the dates when the Rt peak is reached and the associated Rt value. The average Rt value is also reported for the regions that opened schools at 14 September 2020 and the ones that opened schools later at 24 September 2020. Regions that postponed the schools' opening had an average Rt lower than the one registered for regions that opened schools earlier: Rt = 1.27 vs. Rt = 1.46, respectively. The Campania region that applied the most stringent policies in Italy, by closing all schools promptly, registered the lowest Rt value among all the Italian regions. Moreover, we can observe that regions that opened schools earlier had their Rt peaks earlier than regions that postponed the schools' reopening: in the first category (14 September), eight regions had their Rt peak at 10 October 2020 and five regions at 20 October 2020; in the second category (24 September), six regions had their Rt peak at 20 October 2020 and only one region at 10 October, thus suggesting that the impact of school reopening is actually detected after two weeks, as expected.

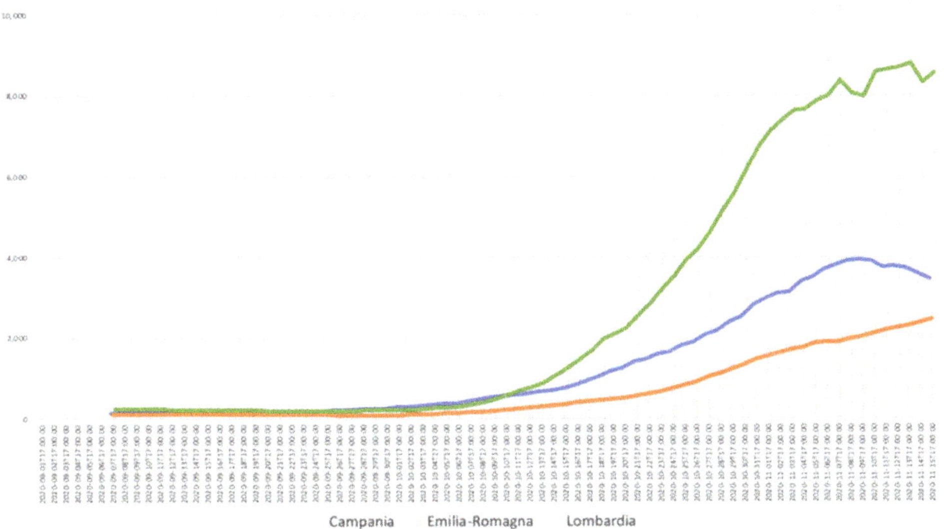

Figure 7. Daily positive new cases and 7-days avg. trend for the Italian regions: Campania, Emilia and Lombardy in the time frame 1 September 2020 to 15 November 2020.

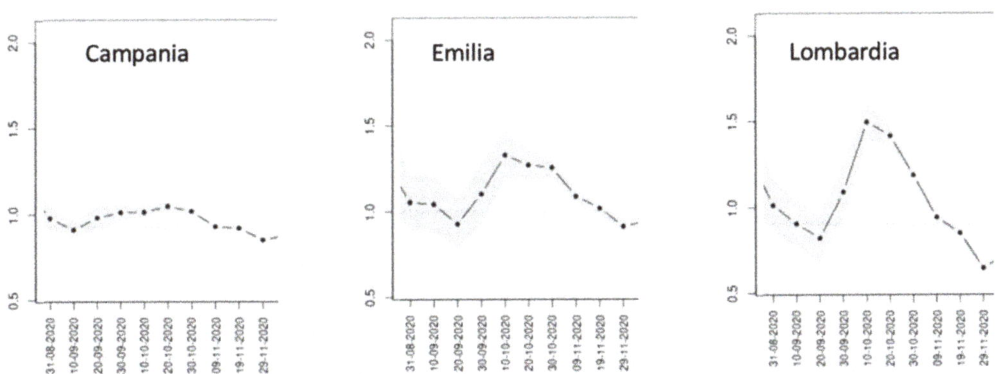

Figure 8. Rt trend for the three Italian regions: Campania, Emilia and Lombardy.

Table 2. Rt peak values and dates associated to Italian Regions.

	14 September 2020			24 September 2020	
	Rt Peak Date	**Rt Peak Value**		**Rt Peak Date**	**Rt Peak Value**
Emilia	10/10/2020	1.4	Abruzzo	10/10/2020	1.3
Lazio	10/20/2020	1.4	Basilicata	10/20/2020	1.4
Liguria	10/10/2020	1.4	Calabria	10/20/2020	1.4
Lombardia	10/10/2020	1.5	Campania	10/20/2020	1.1
Marche	10/20/2020	1.4	Friuli *	10/20/2020	1.3
Molise	10/10/2020	**1.8**	Puglia	10/20/2020	1.2
Piemonte	10/20/2020	1.5	Sardegna *	10/20/2020	1.2
Sicilia	10/10/2020	1.3			
Toscana	10/10/2020	1.4			
Trento	10/20/2020	1.5			
Umbria	10/10/2020	1.4			
Valle d'Aosta	10/10/2020	1.6			
Veneto	10/20/2020	1.4			
Avg.		**1.46**		**Avg.**	**1.27**

* Friuli schools opening 16 September 2020. Sardegna schools opening 22 September 2020.

4. Discussion

Our work shows that the schools' reopening had a clear impact on the overall contagion, since all the three analyzed regions had an increase in the number of cases two weeks after the schools' reopening. The time-lag detected for all the analyzed provinces is equal to 14 days, confirming our hypothesis that two incubation cycles are needed to perceive the impact of the contagion coming from schools. Moreover, the provinces that have had a large number of cases in the school environment are the ones that have subsequently had a higher total number of cases, and as expected, the contagion increased over time. The most significant example is Varese that with reference to the other provinces of the Lombardy Region is the one that had the highest incidence in schools' spreading over time throughout the entire provincial territory, thus leading to one of the most affected provinces in Italy during the second COVID-19 wave. Our study also suggests that population density is another driver of contagion by favoring the virus spread, while the mobility of population (that was already drastically reduced by the governmental restrictions with respect to the normal baseline) did not impact the COVID-19 spread.

Promptly acting by closing the schools (as in the case of RC) was able to contain the COVID-19 spread (i.e., as listed in Table 2, RC was the region with the lowest Rt peak, and it was able to invert the trend of new daily cases before the other two regions RL and REm, as depicted in Figure 7).

It is also interesting to observe, as reported in Table 3, that RL had the highest school contagion index and the highest overall contagion index, while REm had the lowest overall contagion index despite having the same school index of RC. This can be explained by the impact the retail and recreation mobility may have had for RC.

Table 3. Summary contagion findings.

	School Index	Contagion Index	Rt
	(Max-Min-Avg)	(Max-Min-Avg)	(Peak)
Lombardy Region	2.7 0.4 1.3	13.1 2.5 7.4	1.5
Campania Region	1.0 0.2 0.7	9.2 2.6 5.9	1.1
Emilia Region	1.1 0.4 0.7	6.7 3.2 4.5	1.4

5. Conclusions

There are different elements and different factors that suggest us to conclude that the school is not a safe environment by definition, but it must be made sure, by taking serious actions to protect students, teachers, and operators who work and live every day the school context, such as strict personal hygienic conditions, respect for the rules, serious contact tracing activities, timely testing and swabs for students, adequate natural and artificial ventilation of classrooms, etc.

This study may be extended to other Italian regions and to new data, when the MIUR will officially release new data on the infection detected within the schools.

Author Contributions: Conceptualization, D.T. and A.S.C.; Methodology, D.T. and A.S.C.; Software, D.T.; Validation, D.T. and A.S.C.; Formal Analysis, D.T.; Investigation, D.T. and A.S.C.; Data Curation, D.T.; Writing D.T. and A.S.C.; Funding Acquisition, D.T. All authors have read and agreed to the published version of the manuscript.

Funding: This research is partially funded by the ERC Advanced Grant project 693174 GeCo (Data-Driven Genomic Computing), 2016–2021.

Institutional Review Board Statement: Not applicable.

Informed Consent Statement: Not applicable.

Data Availability Statement: Contagion Dataset are available in open format at the GitHub of the Italian Department Civil Protection: https://github.com/pcm-dpc/COVID-19 (accessed on 27 April 2021).

Conflicts of Interest: The authors declare no conflict of interest.

References

1. Tosi, D. Cell Phone Big Data to Compute Mobility Scenarios for Future Smart Cities. *Int. J. Data Sci. Anal.* **2018**, *4*, 265–284. [CrossRef]
2. Tosi, D.; LaRosa, M.; Marzorati, S.; Dondossola, G.; Terruggia, R. Big Data from Cellular Networks: How to Estimate Energy Demand at real-time. In Proceedings of the IEEE International Conference on Data Science and Advanced Analytics (DSAA 2015), Paris, France, 19–21 October 2015.
3. Lavazza, L.; Morasca, S.; Taibi, D.; Tosi, D. Predicting OSS Trustworthiness on the Basis of Elementary Code Assessment. In Proceedings of the ACM/IEEE 4th Empirical Software Engineering and Measurement conference (ESEM 2010), Bolzano, Italy, 16–17 September 2010.
4. Tosi, D.; Chiappa, M. Understanding the Geographical Spread of COVID-19 in relation with Goods Regional Routes and Governmental Decrees: The Lombardy Region Case Study. *Int. J. SN Comput. Sci.* **2020**, *2*, 203. [CrossRef] [PubMed]
5. Tosi, D.; Campi, A. How Data Analytics and Big Data can Help Scientists in Managing COVID-19 Diffusion: A Model to Predict the COVID-19 Diffusion in Italy and Lombardy Region. *Int. J. Med. Internet Res.* **2020**, *22*, e21081. [CrossRef] [PubMed]
6. Tosi, D.; Verde, A. Clarification of Misleading Perceptions of COVID-19 Fatality and Testing Rates in Italy: Data Analysis. *Int. J. Med. Internet Res.* **2020**, *22*, e19825. [CrossRef] [PubMed]

7.   Rennert, L.; McMahan, C.; Kalbaugh, C.A.; Yang, Y.; Lumsden, B.; Dean, D.; Pekarek, L.; Colenda, C.C. Surveillance-based informative testing for detection and containment of SARS-CoV-2 outbreaks on a public university campus: An observational and modelling study. *Lancet Child Adolesc. Health* **2021**. [CrossRef]

8.   Levinson, M.; Cevik, M.; Lipsitch, M. Reopening Primary Schools during the Pandemic. *N. Engl. J. Med.* **2020**, *383*, 981–985. [CrossRef] [PubMed]

9.   Ismail, S.A.; Saliba, V.; Bernal, J.L.; Ramsay, M.E.; Ladhani, S.N. SARS-CoV-2 infection and transmission in educational settings: A prospective, cross-sectional analysis of infection clusters and outbreaks in England. *Lancet Infect. Dis.* **2021**, *21*, 344–353. [CrossRef]

10. Di Domenico, L.; Pullano, G.; Sabbatini, C.E.; Boëlle, P.; Colizza, V. Modelling safe protocols for reopening schools during the COVID-19 pandemic in France. *Nat. Commun.* **2021**, *12*, 1073. [CrossRef] [PubMed]

11. Panovska-Griffiths, J.; Kerr, C.C.; Stuart, R.M.; di Domenico, L.; Pierre-Yves, B.; Vittoria, C. Determining the optimal strategy for reopening schools, the impact of test and trace interventions, and the risk of occurrence of a second COVID-19 epidemic wave in the UK: A modelling study. *Lancet Child Adolesc. Health* **2020**, *4*, 817–827. [CrossRef]

12. Bayham, J.; Fenichel, E.P. Impact of school closures for COVID-19 on the US health-care workforce and net mortality: A modelling study. *Lancet Public Health* **2020**, *5*, e271–e278. [CrossRef]

13. Haug, N.; Geyrhofer, L.; Londei, A.; Dervic, E.; Desvars-Larrive, A.; Loreto, V.; Pinior, B.; Thurner, S.; Klimek, P. Ranking the effectiveness of worldwide COVID-19 government interventions. *Nat. Hum. Behav.* **2020**, *4*, 1303–1312. [CrossRef] [PubMed]

14. Brauner, J.M.; Mindermann, S.; Sharma, M.; Johnston, D.; Salvatier, J.; Gavenčiak, T.; Stephenson, A.B.; Leech, G.; Altman, G.; Mikulik, V.; et al. Inferring the effectiveness of government interventions against COVID-19. *Sci. Dic.* **2020**, *371*, eabd9338.

15. Ferretti, A. Schools and COVID19 Cases in Italy. Available online: https://www.ilfattoquotidiano.it/blog/aferretti/ (accessed on 27 April 2021).

16. Schools and Positive Cases in Lazio Region. Available online: https://roma.repubblica.it/cronaca/2021/03/16/news/scuola_coronavirus_focolai_covid_zona_rossa_roma_lazio-292442424/?ref=fbplrm&fbclid=IwAR1r1yvwvyc57qYug1ZXi5-ewDCvXpRz7i__tmM9WIP72O3pFDsT9jrog4M (accessed on 27 April 2021).

17. Belgium Prime Minister Press Conference (19/03/2021). Available online: https://www.brusselstimes.com/news/belgium-all-news/160897/belgium-consultative-committee-alexander-de-croo-frank-vandenbroucke-outdoor-plan-events-amusement-parks-youth-camps-trains-holiday/ (accessed on 27 April 2021).

18. Press Release COVID19 Cases at Italian Schools. Available online: https://www.facebook.com/prediremegliochecurare (accessed on 27 April 2021).

19. Kar, S.K.; Yasir Arafat, S.M.; Kabir, R.; Sharma, P.; Saxena, S.K. Coping with Mental Health Challenges During COVID-19. In *Medical Virology: From Pathogenesis to Disease Control*; Saxena, S., Ed.; Coronavirus Disease 2019 (COVID-19); Springer: Singapore, 2020. [CrossRef]

20. Pillai, S.; Siddika, N.; Hoque Apu, E.; Kabir, R. COVID-19: Situation of European Countries so Far. *Arch. Med. Res.* **2020**, *51*, 723–725. [CrossRef] [PubMed]

21. Vinnakota, D.; Parsa, A.D.; Arafat, S.Y.; Sivasubramanian, M.; Kabir, R. COVID-19 and risk factors of suicidal behaviour in UK: A content analysis of online newspaper. *J. Affect. Disord. Rep.* **2021**, *4*, 100142. [CrossRef]

22. MIUR Dataset of COVID Infection in Italian Schools. Available online: https://drive.google.com/file/d/1bZuV-UmLd40kxBof1iaPJEJHrlNPUIw3/view (accessed on 27 April 2021).

23. Dashboard National, Regional and Provincial Curves. Available online: www.covid19-italy.it (accessed on 27 April 2021).

24. Official Opendata COVID19 Italy. Department Civil Protection. Available online: https://github.com/pcm-dpc/COVID-19 (accessed on 27 April 2021).

25. Ordinanza Scuole Regione Lombardia. Available online: https://www.regione.lombardia.it/wps/portal/istituzionale/HP/coronavirus/misure (accessed on 27 April 2021).

26. Ordinanza Scuole Regione Campania. Available online: http://www.regioni.it/newsletter/n-3930/del-16-10-2020/campaniaordinanza-di-chiusura-scuole-e-universita-fino-al-30-ottobre-21783/ (accessed on 27 April 2021).

27. Covid19 Google Community Mobility Report. Available online: https://www.google.com/covid19/mobility/ (accessed on 27 April 2021).

28. Wallinga, J.; Teunis, P. Different epidemic curves for severe acute respiratory syndrome reveal similar impacts of control measures. *Am. J. Epidemiol.* **2004**, *160*, 509–516. [CrossRef] [PubMed]

future internet

MDPI

Article

Multi-Attribute Decision Making for Energy-Efficient Public Transport Network Selection in Smart Cities

Rashmi Munjal *, William Liu, Xuejun Li, Jairo Gutierrez and Peter Han Joo Chong *

School of Engineering, Computer, and Mathematical Sciences, Auckland University of Technology, Auckland 1142, New Zealand; william.liu@aut.ac.nz (W.L.); xuejun.li@aut.ac.nz (X.L.); jairo.gutierrez@aut.ac.nz (J.G.)
* Correspondence: rashmi.munjal@aut.ac.nz (R.M.); peter.chong@aut.ac.nz (P.H.J.C.)

Abstract: Smart cities use many smart devices to facilitate the well-being of society by different means. However, these smart devices create great challenges, such as energy consumption and carbon emissions. The proposed research lies in communication technologies to deal with big data-driven applications. Aiming at multiple sources of big data in a smart city, we propose a public transport-assisted data-dissemination system to utilize public transport as another communication medium, along with other networks, with the help of software-defined technology. Our main objective is to minimize energy consumption with the maximum delivery of data. A multi-attribute decision-making strategy is adopted for the selction of the best network among wired, wireless, and public transport networks, based upon users' requirements and different services. Once public transport is selected as the best network, the Capacitated Vehicle Routing Problem (CVRP) will be implemented to offload data onto buses as per the maximum capacity of buses. For validation, the case of Auckland Transport is used to offload data onto buses for energy-efficient delay-tolerant data transmission. Experimental results show that buses can be utilized efficiently to deliver data as per their demands and consume 33% less energy in comparison to other networks.

Keywords: big data; delay-tolerant network (DTN); multi-attribute decision making; public transport; energy consumption

Citation: Munjal, R.; Liu, W.; Li, X.; Gutierrez, J.; Chong, P.H.J. Multi-Attribute Decision Making for Energy-Efficient Public Transport Network Selection in Smart Cities. *Future Internet* **2022**, *14*, 42. https://doi.org/10.3390/fi14020042

Academic Editor: Davide Tosi

Received: 23 December 2021
Accepted: 20 January 2022
Published: 26 January 2022

Publisher's Note: MDPI stays neutral with regard to jurisdictional claims in published maps and institutional affiliations.

1. Introduction

The smart city is being equipped with many smart devices, driven by the advancement of digital technologies and the ever-increasing demand of end-user applications. However, energy-efficiency is one of the recent demands toward the development of the green smart city. It is estimated that smart cities will be equipped with possibly 40,000 million smart devices for 100,000 million global connections in different areas, such as health care, transportation, and finance, etc. These smart devices will be responsible for generating big data in the smart city, which is already increasing at a compound annual growth rate (CAGR) of 47%. It has been estimated that 90 ZB of data will be created on IoT devices by 2025 [1].

As companies currently transfer massive amounts of data across wide-area networks to backup their data, sync search indexes between data centers, or provide high-definition surveillance video records to governments and access audio and video across social media sites, a large amount of data is transferred over wide-area networks. Since the data volume and complexity of big data [2] are extremely large, the survival of big data is impossible without the underlying technical support of networking. A new connectivity method is, therefore, required to overcome this biggest challenge. By finding alternative data-transmission network architectures, researchers aim to reduce traffic congestion. Cellular base stations, T2T approaches, WI-FI hotspots, and vehicular networks are a few examples of data offloading techniques used. Cities' bus networks [3] have characteristics such as wide coverage and fixed routes, granting them the potential to form the backbone of communication, alongside traditional networks.

In recent years, vehicular-assisted networks hold the utmost importance in the smart city to improve the quality of life, reliability, operational efficiency, and service quality in urban areas. Vehicles are used as data carriers in network communication. In addition to this, the data-offloading approach has been utilized to offload data from one network to another as per different criteria and priorities. Mobile Computation Offloading (MCO) [4] is a popular emerging technology to offload computation-intensive data to the servers to increase the capacity of devices and conserve battery energy. Through opportunistic contacts between moving vehicles and Road-Side Units (RSUs) placed on roads, it is possible to offload data onto vehicles for further delivery. In particular, public transport is a category of vehicular networks with several exclusive properties, such as regular and scheduled movements and reliable physical coverage in all urban centers.

The main contributions of this paper are the following:

1. We designed the Public Transport-Assisted Data-Dissemination (PTDD) System in a smart city which will be equipped with wireless sensors and data centers to handle massive data using wired, wireless, and public transport networks;
2. We applied a Multi-Attribute Decision making (MADM) algorithm for best network selection based upon different user requirements and different attributes;
3. We applied the Capacitated Vehicle Routing Problem (CVRP) to minimize energy consumption using public transport as a data carrier. We will use buses to offload the entire set of demands of each bus stop. Our model constrains the objective by the maximum capacity of the bus;
4. For the evaluation of the best network selection, different services are considered, based upon user requirements, to find the best network in the heterogeneous network. Next, a detailed comparative analysis of energy consumption is performed for traditional and public transport networks for the various demands of users.

The rest of the paper is organized as follows. The PTDD is presented in Section 2, along with MADM and CVRP, for network selection and for allocating data onto buses. In Section 3, we perform a numerical analysis and include two case studies to present the results. Next, we have a brief discussion section in Section 4. Finally, the paper is concluded in Section 5, along with a brief discussion about future work.

2. Related Work

Energy-efficient network technology is defined as the better utilization of resources whenever possible to alleviate network congestion. It has been estimated that 3% of the world's yearly electrical energy consumption, and 2% of CO_2 emissions, are caused by information and communication technology (ICT) infrastructure [5]. Moreover, it is estimated that ICT energy consumption [6,7] is rising by 15–20 percent per year. Specifically, 57% of the energy consumption of the ICT business goes to users and network devices on mobile and remote networks [8]. The rapid development of energy consumption by the user and network devices has created major issues [9]; many efforts are being made by researchers for sustaining quality of services, throughput, and adaptability [10,11]. Devices, and their infrastructures, are arranged to obtain good QoS, and to provide better utilization of resources. The trade-off between execution and energy utilization should be exploited. The connection between energy and execution is indicated by [8]. The goal of energy efficiency is achieved through the use of virtualization, the consolidation of servers, and by upgrading older products to new, more energy-efficient ones.

Many co-operative data collection approaches from different locations have been proposed [12]. These approaches find [13,14] vehicles as optimal and logical links for transferring big data. Therefore, traditional homogeneous network communication, hand-off algorithms, and data offloading are a few diverse applications [15] proposed to offload data onto different networks while considering different attributes. The public transport-assisted data-dissemination system can be interpreted as a delay-tolerant network where RSUs will be placed on each bus stop and communication between buses and RSUs occurs once the bus stops or passes by bus stops.

Before offloading data onto these buses, network selection is the critical process of identifying the best network in a heterogeneous network. This is possible with Multi-Attribute Decision-Making (MADM) algorithms for appropriate decisions among different networks. There is a vast literature on MADM-based network-selection algorithms [16–18]. Many of these studies are user-centric and help to make decisions based on user preferences. There are many MADM algorithms for solving the network-selection problem, including AHP, GRA, SAW, MEW, TOPSIS, DIA, and ELECTRE [19]. Many researchers have discovered many other types of algorithms to resolve VHO and network-selection problems in heterogeneous networks. Some of them are utility functions [20], genetic algorithms [21], or use game theory principles [22]. Utility functions assign values as per the ranking of choices for the user's satisfaction. Abid et al. [23] proposed an innovative single-criteria utility function that used a metric for measuring user satisfaction as well as sensitivity to each decision criterion for deciding whether to hand over.

In [24], the researchers proposed a utility-based fuzzy-Analytic Hierarchy Process (AHP)-based network selection in heterogeneous wireless networks. They categorized different applications, such as voice, video, and best effort, and triangular fuzzy numbers were used to represent their comparison matrices. The results obtained prove that the MEW method yields better scores with utility functions. Jiang et al. [25] proposed a joint multi-criteria utility-based algorithm to assist the vehicle in infrastructure networking for energy efficiency. A user's preferences for different attributes, such as bandwidth, delay, signal intensity, and network cost, help to establish utility functions and an energy-efficient network-selection algorithm. Additionally, there have been some papers published on energy-efficient multi-connection for 5G heterogeneous networks [26].

Michele et al. [27] explored the BUSNET algorithm that achieves effective routing in a bus environment. It considers routing at a bus-line level instead of a bus level. ALARMS [28] is one of the message-scheduling approaches that uses message ferries to forward messages and achieve good QoS. This publication [29] gives a promising solution, namely, "Cost-Effective Multimode Offloading"(CEMMO), that offloads data to the best possible choice among the following three options to reduce the overall cost in terms of energy efficiency, financial settlement, and user satisfaction. Kessar et al. [30] introduced the Always Best Connected (ABC) concept for always providing the best connectivity to all the applications. The handover decision is being taken on regrouping criteria such as network, terminal, user, and services. Another network-selection mechanism [31] was used in combination with AHP and GRA to trade off network circumstances, services, and user priorities. AHP was used for weighing based upon user preferences and GRA was used for ranking network alternatives. Liang et al. [32] introduced a user-oriented network-selection scheme, where five different modules are considered for network selection. One of them is an input which includes a utility function, and the other is a user-preference calculation using FAHP to calculate weighing of judgment. Yu et al. [33] proposed network selection using multi-service multi-modal terminals. They also used utility functions for multi-services for user satisfaction, network attributes, and service characteristics. In our previous work [3,34,35], we have introduced the use of a public transport network and offloaded data onto buses along with other networks for energy efficiency. We extended our work in the proposed manuscript with network selection and appropriate vehicle selection to offload data for energy efficiency.

3. Public Transport-Assisted Data-Dissemination System

The proposed framework depicts the Public Transport-Assisted Data-Dissemination System (PTDD), which consist of smart cities that are equipped with wireless sensors and data centers to handle massive data dissemination for several categories of applications, as shown in Figure 1, using a set of buses picked up at each bus stop. PTDD is composed of a central controller and a data center, along with RSUs deployed at bus stops and onboard units on buses. Smart meters, video surveillance data, and air pollution data are some of the delay-tolerant applications and can tolerate delays ranging from seconds, to minutes, to hours.

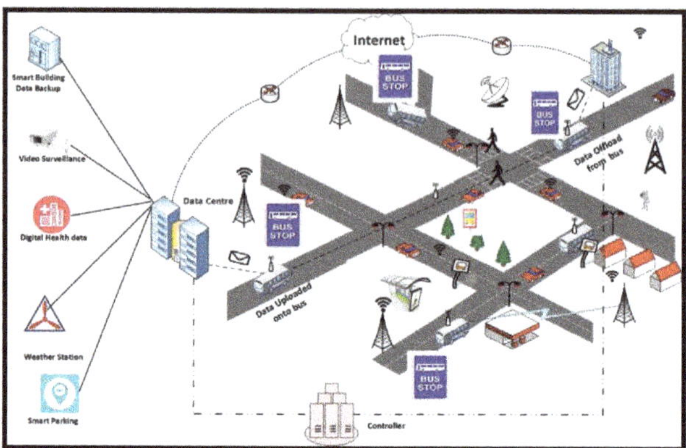

Figure 1. Public Transport-Assisted Data-Dissemination System (PTDD)

Over the last few years, we have witnessed the rapid growth of vehicles in urban areas together with the increase of internet-enabled devices integrated into vehicles [36]. Vehicles are being used as mobile nodes to create a mobile ad hoc network. They move randomly and communicate either with moving vehicles or fixed equipment such as RSUs. This alternative communications network layer of public transport networks will include public vehicles moving around the city. The flowchart given in Figure 2 gives an overview of the overall workflow of the proposed system. We will first apply the MADM methodology for the selction of the best network in the heterogeneous network, and next, we will offload data onto selected public vehicles to carry it until the destination for energy-efficient data transmission.

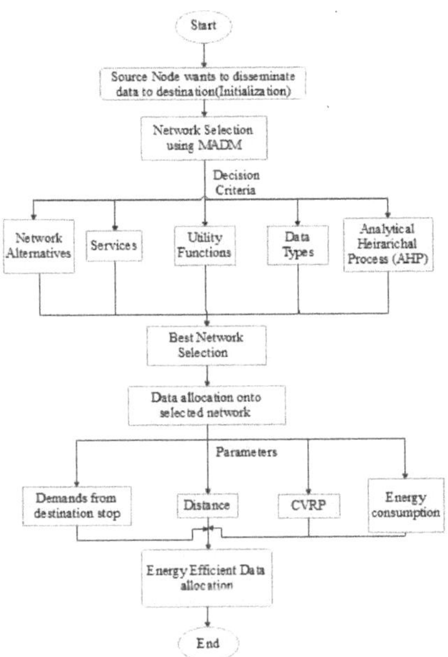

Figure 2. Flowchart of the proposed data-dissemination system.

3.1. Multi-Attribute Decision Making

MADM is being used for network selection among all the available networks. The network-selection procedure ultimately aims for the best network that can support the required service(s) and avoid excessive switching among different networks to minimize service interruptions and energy consumption. Therefore, we introduce the MADM method used by the controller in response to suitable network selection. This model helps to make forwarding decisions fairly. MADM is an important tool that assists in the solution of complex decision-making problems and analyzes network-selection problems in a heterogeneous network. There are a few characteristics of MADM given below:

(a)   Alternatives: Alternatives are defined as several different options to prioritize or select. These can be called candidates, users, or networks, etc.;

(b)   Decision Matrix: Any MADM problem can be mathematically defined by using a decision matrix, $L(M \times N)$:

$$
L = \begin{array}{c}
\begin{array}{ccccccc}
C_1 & C_2 & \cdots & C_j & \cdots & C_N &
\end{array} \\
\begin{pmatrix}
x_{1,1} & x_{1,2} & \cdots & x_{1,j} & \cdots & x_{1,N} \\
x_{2,1} & x_{2,2} & \cdots & x_{2,j} & \cdots & x_{2,N} \\
\vdots & \vdots & \ddots & \vdots & \ddots & \vdots \\
x_{i,1} & x_{i,2} & \cdots & x_{i,j} & \cdots & x_{i,N} \\
\vdots & \vdots & \ddots & \vdots & \ddots & \vdots \\
x_{M,1} & x_{M,2} & \cdots & x_{M,j} & \cdots & x_{M,N}
\end{pmatrix}
\begin{array}{c}
A_1 \\ A_2 \\ \vdots \\ A_i \\ \vdots \\ A_M
\end{array}
\end{array} , \tag{1}
$$

where $A_1, A_2, A_3, \ldots .A_i, \ldots ., A_M$ denotes all the alternatives/parameters to consider for decision making. $C_1, C_2, C_3, \ldots .C_j, \ldots ., C_N$ represents all N criteria, which is being calculated as per different alternatives and denotes its performance. For example, $x_{i,j}$ is the performance ranking of the *ith* alternative w.r.t. to the *jth* alternative. The main aim of the decision matrix is to select the best alternative from the given alternatives with respect to others;

(c)   Attribute Weight: Attribute weight is the value obtained by the decision-maker as per each attribute of the network. This weight depends upon the value assigned to the attribute. This weight is calculated by the pairwise comparison matrix;

(d)   Normalization: The attribute used for network selection has different measurement units. Therefore, normalization is a necessary step for this calculation.

MADM algorithms have high accuracy and low difficulty. They capture different parameters (e.g., QoS, bandwidth, delay, data volume, cost, etc.) and select the most suitable network. There are many possible solutions for MADM problems. The whole process of network selection is shown below, in Figure 3.

3.1.1. Initialization Step

The initialization step is the first-most step of the MADM process, which gathers the required information and triggers the process. In this step, there are the following options to consider:

- Service's Requirement: The most important aspect is the user's requirements. For different users, they have different demands and objectives. In our proposed system, we categorize users' requirements into three categories, such as Service 1, Service 2, and Service 3. Different services have different levels of sensitivity to the same networking attribute. For example, considering bandwidth as an attribute, if its service 1, a lower bandwidth will be used. However, if it is a large data transfer, a higher bandwidth will be used. In addition to that, it is assumed that a user can select any one service at one time. Users can select the priority of services used. They can select the urgency or non-urgency of data delivery, which relates to the data type, such as delay-tolerant or delay-sensitive, and helps the controller to make optimal network-selection decisions;
- Data Type: Data types belong to the type of application selected by users. It can be delay-tolerant or delay-sensitive. Some of the services, such as video or data type, can be categorized as a real-time or non-real-time application and can, accordingly, be delayed for some time. This is another important piece of information to consider for optimal network selection;
- Network Alternatives: In our proposed work, we are demonstrating the offloading of data from traditional networks to road networks with delay-tolerant conditions. Therefore, to choose among a list of networks, we will be considering WLAN, UMTS, and Vehicular Networks. The controller will choose the best optimal network among these networks based upon user requirements and data type. Three of these networks have different properties. The vehicular network is used for all delay-tolerant applications, such as emails, data backup, video download, and photos, which significantly contribute to energy efficiency without a negative effect on user satisfaction. We assume that all vehicles are equipped with On-Board Units (OBU) to carry data. If we compare the other two networks, WLAN networks are managed for higher bandwidths and lower delay applications, although UMTS networks are the most energy-efficient with lower bandwidth requirements and large delays.

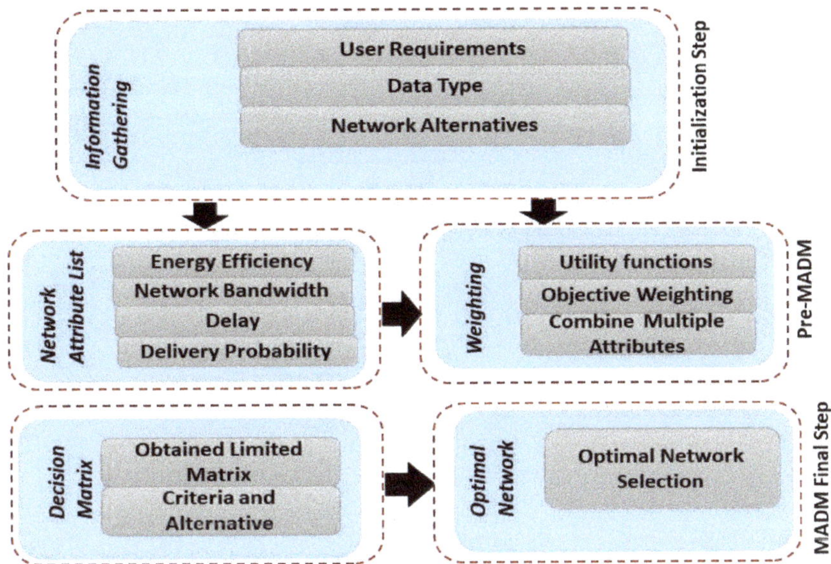

Figure 3. MADM for network selection.

Next, considering all the requirements in the process of network selection, we integrate utility theory with the AHP process to design our network-selection algorithm, as shown in Figure 4. We consider the characteristics of different types of services and their respective weights to define utility functions and the scores of a user's preferences by defining rank preference through AHP. Therefore, we are providing a comprehensive structure for users to give their preference, which the controller can use to make decisions based upon their requirements.

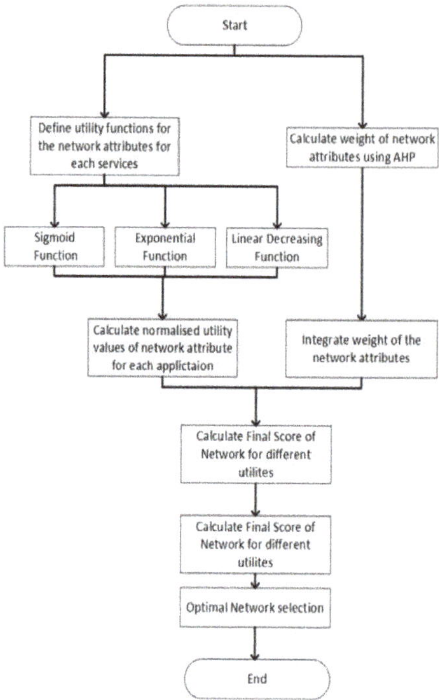

Figure 4. Flow chart of the proposed network-selection algorithm.

3.1.2. Pre-MADM

This step includes preparations before combining all the criteria, including the weighting and the attributes' adjustment procedures. The left part of this step is more about defining all the attributes to decide on the optimal network. The network attribute list consists of energy efficiency, delay-tolerant value, network bandwidth, and delivery probability. The right part of this step assigns utility values for each attribute, weighs different attributes against each other, and gives the best permutation to analyze optimal network selection. In our proposed method, users decide on all the requirements and importances. The controller collects these requirements and proceeds further with the weighing procedure. The measurement metrics for energy efficiency, delivery probability, network bandwidth, and delay are determined by these parameters appropriately.

- Utility function—theory-based network:
  Utility functions measure the level of satisfaction for each user as per different attributes of each network alternative. We design utility functions to map decision factors to the respective utility metrics in order to evaluate the decision factors of network selection. We consider user requirements as per their profile, delay-tolerant indicator (DTI), both network properties, and QoS requirements. There are generally three types of utility functions that network selection uses: (1) sigmoid; (2) monotonically increasing; (3) linearly decreasing. These functions are further categorized as beneficial or non-beneficial criteria. The sigmoid utility function is used with given minimum and maximum requirements. Bandwidth and energy efficiency are beneficial criteria and can be represented as a sigmoid function. The utility theory states that utility functions must satisfy twice differentiability, monotonicity, and concavity–convexity [37]. Therefore, we design different utility functions for different objectives. The value of the utility function lies between 0 and 1. For the most satisfied user, it is 1, and for the least satisfied user, it counts as 0.

- Utility function for Energy Efficiency EE: In this utility function, EE, as discussed, is a beneficial criterion, and the energy-efficient utility function will be modeled as a sigmoid curve. The sigmoidal utility function is defined below:

$$u(e) = \frac{1}{1 + xe^{c(e_{avg}-e)}}; e > 0,\tag{2}$$

where e_{avg} and e represent the average network energy efficiency and network energy efficiency; x is used as a constant value that is always greater than zero $(x > 0)$. The notation c is used to denote the sensitivity of network attributes affecting energy efficiency. The utility function for EE is plotted in Figure 5; we can make sure that the utility function is monotonic and concave–convex. In physical terms, Equation (2) is the result of a higher network energy efficiency, with e translating into a larger utility function, $u(e)$, resulting in a more preferred network.

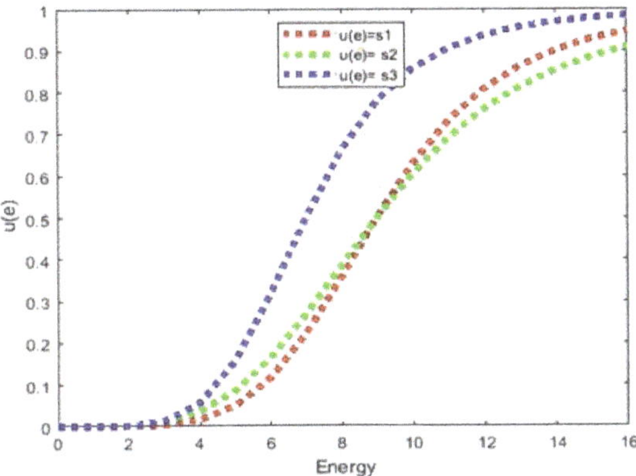

Figure 5. The utility function for energy efficiency.

- Utility function for Network Bandwidth: Network Bandwidth is an important attribute for network selection. For three of these networks, the network bandwidth has a different value. When the network bandwidth is lower than the required bandwidth, as per different service requirements, then there is a compromise in QoS, and there will be a loss of packets. We are using the following utility function to define bandwidth requirements for different applications:

$$u(b) = \begin{cases} 0, & ; b < b_{min} \\ \dfrac{(\frac{b}{b_{med}})^{x4}}{1+(\frac{b}{b_{med}})^{x4}} & ; b \leq b_{min} \leq b_{med} \\ 1 - \dfrac{(\frac{b_{max}-b}{b_{max}-b_{med}})^{x4}}{1+(\frac{b_{max}-b}{b_{max}-b_{med}})^{x4}} & ; b_{med} \leq b \leq b_{max} \\ 1 & ; b > b_{max}, \end{cases}\tag{3}$$

where b_{min} and b_{max} define the minimum and maximum bandwidths of each network. In addition, b is the actual bandwidth required by the user, as per the services required. This is the same as an energy utility function. All the utility functions fulfill the conditions of being is twice differentiable, monotonic, and concave–convex, as shown in Figure 6.

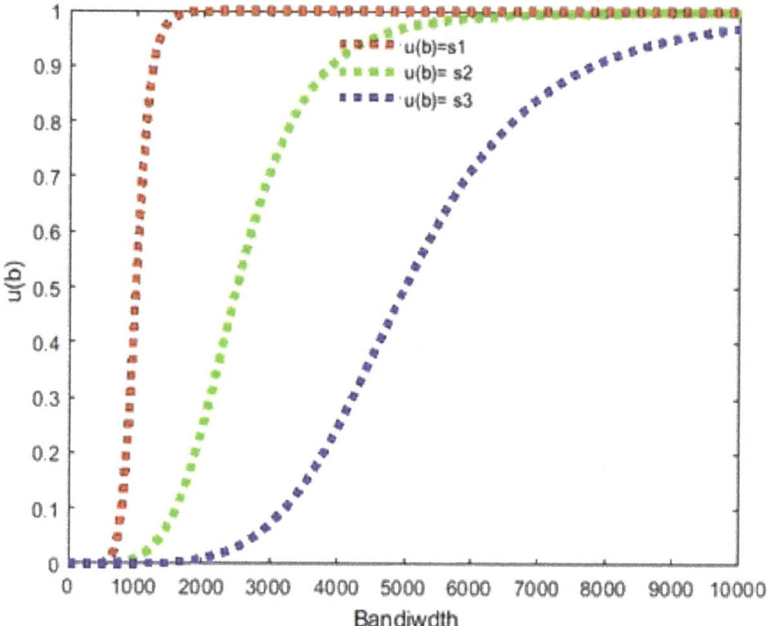

Figure 6. The utility function for Bandwidth.

- Utility function for Delay Tolerance: Generally, incremental latency values are acceptable in a Delay-Tolerant Networks (DTN). While designing the utility function for network delay tolerance, a larger network delay value will result in a lower utility value. It is a decreasing criterion to measure network delay. Delay varies in both networks as per the data volume. u(d) is defined as a utility function for the delay, as below:

$$u(d) = 1 - u'(d) \tag{4}$$

$$u'(d) = \begin{cases} \dfrac{(\frac{d}{d_{med}})^{x3}}{1+(\frac{d}{d_{med}})^{x3}} & ; d \leq d_{min} \leq d_{med} \\[2ex] 1 - \dfrac{(\frac{d_{max}-d}{d_{max}-d_{med}})^{x3}}{1+(\frac{d_{max}-d}{d_{max}-d_{med}})^{x3}} & ; d_{med} \leq d \leq d_{max} \\[2ex] 1 & ; d > d_{max}, \end{cases} \tag{5}$$

where d_{max} is the maximum delay and x is the sensitivity factor for delay calculation among both networks. The delay utility function is shown in Figure 7.

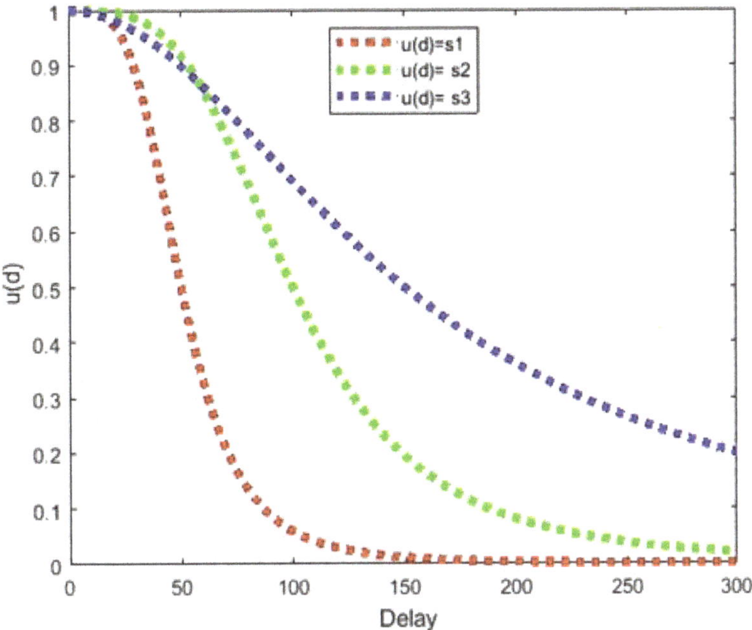

Figure 7. The utility function for delay tolerance.

- Utility function for the Delivery Probability: Delivery probability is to be defined as the volume of data to be sent using any of the networks. We defined the utility function of delivery probability as $u(dp)$, where $dp\epsilon[0,1]$, in case of successful delivery, is 1, and otherwise, for packet loss, it will be considered as 0. Otherwise, it lies between 0 and 1. dp is the delivery probability obtained and dp_{max} is the maximum delivery probability that is acceptable to the user, and is shown in Figure 8.

$$u(dp) = \begin{cases} \frac{dp}{dp_{max}} & ;0 \leq dp \leq dp_{max} \\ 1 & ;dp > dp_{max} \end{cases} \tag{6}$$

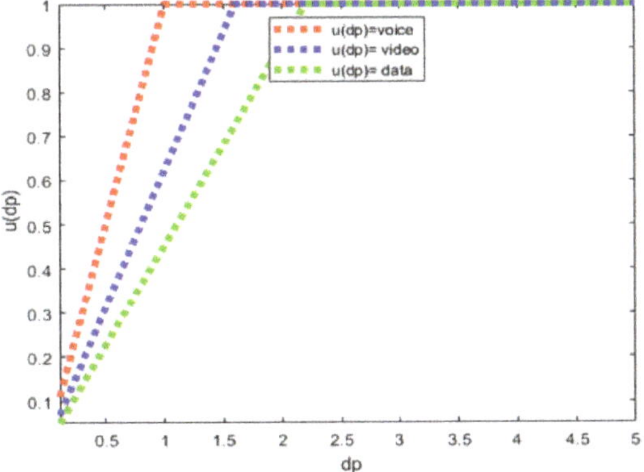

Figure 8. The utility function for delivery probability.

3.1.3. MADM

This step helps with deciding between different networks, based on the weights obtained from the decision matrix, the alternatives, and different criteria.

- Analytical Hierarchical Process
  The analytical Hierarchical process (AHP) method is a multi-criteria decision-making process for network selection. It was developed at the Wharton School of Business by Thomas Saaty in the 1970s [38]. AHP works on the function of priority and rank to evaluate subjective weights to achieve the specified goals. We have used this process to select a best-featured network from the given alternatives for the given service class based on the following criteria—Energy Consumption, Bandwidth, Delay, and Delivery Probability. We have also used this process for choosing a priority of network types for each data type. Network weighing is an important factor to characterize the network performance and user's preferences. We use the hierarchy analysis method to allocate the appropriate weight to each selection metric.
  We further categorize traditional networks into WLAN and UMTS networks for impartial scheming with different attributes, as shown in Figure 9. The logical flowchart of the AHP algorithm considers the hierarchical structure with the main goal, multiple criteria, and network alternatives to select. We have defined utility functions for all the attributes for a network assessment. A user's preference will be based on multiple criteria for network selection. We assume that WLAN users have wireless access to their system, but with a fixed location—or we can say a local network—and that they use all their devices to avail the services and disseminate data to nearby RSUs for further transmission. However, they have good speed and bandwidth values. On the other hand, UMTS is a mobile cellular device and can roam around with their data plans, but with limited bandwidths and larger delays as per the delivery probability and data network's range.

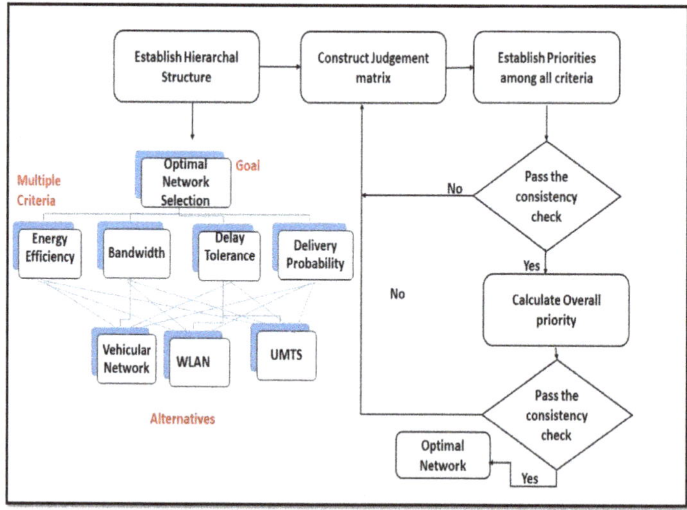

Figure 9. AHP for network selection.

1. Subdivide a problem into further sub-problems by defining an objective function, criteria, and possible alternatives. Here, the objective is our goal of achieving optimal network selection. The multiple criteria are the factors affecting the preference for selection.
2. Develop the hierarchy model of all objectives along with their elements to obtain the priorities of criteria through pairwise comparison matrices.

3. Construct a pairwise comparison matrix for each criterion of hierarchical structure in such a way that all associated criteria are compared with each other as per the intensity of importance [39], with respect to the scale. We believe that a pairwise comparison between alternatives helps for qualitative judgment. This qualitative pairwise comparison follows the importance scale, as shown in Table 1.

$$
P = \begin{array}{c}
\begin{array}{cccccc} C_1 & C_2 & \cdots & C_j & \cdots & C_N \end{array} \\
\begin{pmatrix}
1 & x_{1,2} & \cdots & x_{1,j} & \cdots & x_{1,N} \\
x_{2,1} & 1 & \cdots & x_{2,j} & \cdots & x_{2,N} \\
\vdots & \vdots & 1 & \vdots & \ddots & \vdots \\
x_{i,1} & x_{i,2} & \cdots & 1 & \cdots & x_{i,N} \\
\vdots & \vdots & \ddots & \vdots & 1 & \vdots \\
x_{M,1} & x_{M,2} & \cdots & x_{M,j} & \cdots & 1
\end{pmatrix}
\begin{array}{c} C_1 \\ C_2 \\ \vdots \\ C_i \\ \vdots \\ C_N \end{array}
\end{array}
\tag{7}
$$

Table 1. Criteria importance scale in a pairwise comparison.

Preferences as per Importance	Definition
1	Equal Importance
3	Moderate importance
5	Strong importance
7	Very strong importance
9	Extreme importance
$2, 4, \dots, 8$	Intermediate values

4. Perform the normalization of a given matrix P, which is now denoted as P_{Norm}:

$$
P_{Norm} = \begin{array}{c}
\begin{array}{cccccc} C_1 & C_2 & \cdots & C_j & \cdots & C_N \end{array} \\
\begin{pmatrix}
1 & z_{1,2} & \cdots & z_{1,j} & \cdots & z_{1,N} \\
z_{2,1} & 1 & \cdots & z_{2,j} & \cdots & z_{2,N} \\
\vdots & \vdots & 1 & \vdots & \ddots & \vdots \\
z_{i,1} & z_{i,2} & \cdots & 1 & \cdots & z_{i,N} \\
\vdots & \vdots & \ddots & \vdots & 1 & \vdots \\
z_{M,1} & z_{M,2} & \cdots & z_{M,j} & \cdots & 1
\end{pmatrix}
\begin{array}{c} C_1 \\ C_2 \\ \vdots \\ C_i \\ \vdots \\ C_N \end{array}
\end{array}
\tag{8}
$$

$$
\text{where, } z_{i,j} = \frac{x_{i,j}}{\sum_{i=1}^{N} x_{i,j}}. \tag{9}
$$

5. The contributions of each normalized metric are multiplied by the assigned importance weight wj, and can be calculated for the ith criteria, as below:

$$
P_w = \frac{\sum_{i=1}^{N} Z_{i,j}}{N} \text{ with } \sum_{i=1}^{N} P_w = 1, \tag{10}
$$

such that P_w is the weight vector.

6. Calculate the consistency index, where λ_{max} is the largest eigenvalue of P_{Norm}, and it is determined from the eigenvalue computation of P_{Norm}:

$$
CI = \frac{\lambda_{max} - N}{N - 1}. \tag{11}
$$

7. In the last step, evaluate the consistency of the comparison using the Consistency Ratio (CR), defined as:

$$CR = \frac{CI}{RI},\tag{12}$$

where RI [31], as defined in Table 2, is the index used for the number of attributes used in decision making; the network is ranked based on this index. For acceptable results, $CR < 0.1$; otherwise, pairwise comparison should be repeated.

Table 2. The random index.

N	1	2	3	4	5	6	7	8	9	10
RI	0	0	0.58	0.9	1.12	1.24	1.32	1.41	1.45	1.49

In such a way, AHP helps with network selection among different networks based upon different attributes. After the selection of the public transport network, the next section will elaborate further about allocating data onto buses as per their stay-time at each bus stop.

3.2. Capacitated Vehicle Routing Problem (CVRP)

Once we select the best network in terms of energy efficiency. It is important to know which vehicle can be more energy-efficient when we allocate data onto buses at each bus stop. As shown below, in Figure 10, the source data center accumulates all the data from nearby user devices and caches it until an optimal bus is not found for the destination route. At each bus stop, RSUs have been deployed to offload data onto buses, and these buses carry data until the destination bus stop and upload onto the destination bus stop and to the data center.

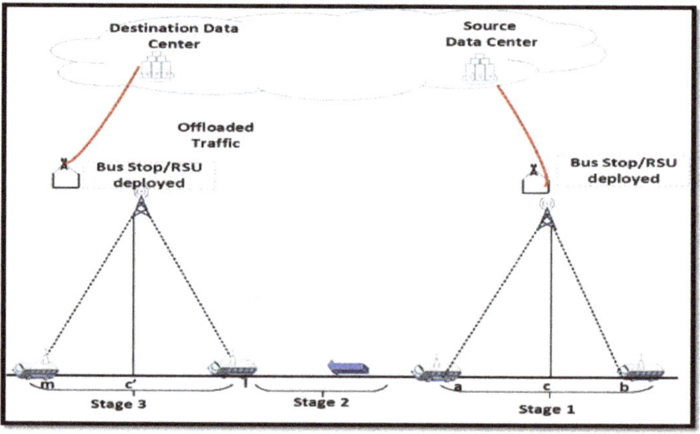

Figure 10. CVRP problem for data allocation.

We will take different demands from the data center (DC) to allocate data onto buses as per their maximum capacity to carry data until reaching the destination while minimizing energy consumption. We will, first, define CVRP to minimize energy consumption while using public transport as a data carrier.

In our model, all demands are allocated by the controller to the appropriate bus going in that direction. The demands are fetched from the DC and are allocated to the bus going on a trip in the direction of the destination location. Note that data offloading/uploading is possible at each bus stop; therefore, the transmission range is expected to be limited for data offloading onto these buses. The whole transmission procedure and energy consumption is calculated in three stages:

Stage 1: RSU–Bus transmission: When the bus stops at the parent stop or source data center, data is allocated onto the bus within the transmission range. As shown in Figure 10, a and b are the earliest and final points for stage 1. Point c denotes the central projection when the bus stops at the bus stop;

Stage 2: Stable State: In stage 2, the bus will carry data, as per demands, on its fixed route and does not consume any extra energy, and will consume negligible energy.

Stage 3: Bus–RSU transmission: In stage 3, the bus reaches the destination spot and uploads data onto the bus stop. l and m are the initial and final points of this stage, and c' is the vertical projection of RSU deployed at the destination bus stop. We will minimize energy consumption by offloading data onto the fixed bus with a fixed capacity to carry data and, thus, finding the optimal solution.

- Problem Definition

  To offload data onto buses, there is n number of demands being fulfilled by a DC, and a nearby stop is a depot to start the bus journey and return to the same bus stop after finishing its route. B is the set of buses, CB is the capacity of the bus, D is the deadline for the message delivery, which also considers the number of trips being taken by a bus. Each DC has different demands di for different locations. We define our problem in a graph $G(V, E)$, where $V = 0, 1, 2 \dots n$ is a set of all nodes of the graph and E is the set of edges $(i, j) \dots (I, j) \epsilon N$. Arc (i, j) represents the path from node i to node j. The energy cost $(E_{i,j})$ is calculated for each bus to carry data from the source until the destination. The minimum number of buses required to fulfill all the demands is $\frac{\sum_{i=1}^{n} d_i}{C_B}$. The controller will assign demands onto each bus as per the destination location. A CVRP can be formulated as follows:

Objective: To minimize

$$\sum_{b \in B} \sum_{i=1}^{n} \sum_{j=1}^{n} E_{i,j} X_{i,j,b}, \tag{13}$$

which minimizes the total energy consumption cost of buses. There are various constraints subjected to this function, defined below:

Subjected to:

$$\sum_{i=1, i \neq j}^{n} \sum_{b \epsilon B} X_{b,i,j} = 1 \qquad \forall j = 1, \dots n \tag{14}$$

$$\sum_{j=1}^{n} X_{b,0,j} = 1 \qquad \forall \, b \epsilon (B_1, B_2, \dots B_n) \tag{15}$$

$$\sum_{i=1, i \neq j}^{n} X_{b,i,j} = \sum_{i=1}^{n} X_{b,i,j} \qquad \forall j = 1, \dots n, \quad b \epsilon (B_1, B_2, \dots B_n) \tag{16}$$

$$\sum_{i=1}^{n} \sum_{j=1, i \neq j}^{n} d_j X_{b,i,j} \leq C_B \qquad \forall b \epsilon (B_1, B_2, \dots B_n) \tag{17}$$

$$\sum_{b=B_1}^{B_n} \sum_{i \epsilon T} \sum_{j \epsilon T, i \neq j} X_{b,i,j} \leq |T| - 1 \qquad \forall T \subseteq (1, \dots n) \tag{18}$$

$$X_{b,i,j} \epsilon (0, 1) \qquad \forall b \epsilon (B_1, B_2, \dots B_n); i, j = (1, \dots n) \tag{19}$$

where $X_{(i,j,b)}$, the binary variable, defines a set of buses $b \epsilon B_1, B_2 \dots B_n$, that traverses an arc (i, j). The objective function, defined in equation 13, minimizes the energy-consumption cost. Constraint 14 is the degree constraints, confirming that each demand will be fulfilled by an available bus. Each bus starts its trip from the parent stop, where data is offloaded, delivers data at the destination, and finishes the trip at the same stop as shown in constraint 15 and 16. Constraint 17 defines the maximum capacity of the bus to carry data. All the demands of the DC are fulfilled by the available buses of the day. Constraint 18 defines that, as per the defined time, there are no cycles disconnected to the parent stop. The definition domains of the variables are described in constraint 19.

4. Numerical Analysis and Results

Firstly, to evaluate the best network selection, we will consider the Auckland public transport network to choose among three network alternatives. The reason for choosing Auckland as a case study is that Auckland has a vision to be the world's most liveable city with smart citizens and a smart infrastructure. Auckland is a city with innovative technologies to improve the quality of life. Auckland, as a smart city, can think of smart and innovative devices to make decisions based on real-time data analysis. Seven New Zealand projects have been short-listed in IDC's Asia Pacific Smart Cities Awards [40]. Considering all the facts, Auckland Transport was a good example to validate our proposed system. Normally, the urban area is covered by heterogeneous wireless networks, including WLAN, UMTS, and Public Vehicles/buses. All these networks bear different characteristics, as described above. For vehicular network selection, the vehicle must be in the range of the network to consider it a selection option, based upon the user's preference. For simplification, we make the following assumptions:

Assumption 1. *We consider three types of networks: WLAN, UMTS, and vehicular networks. For further information related to the vehicular network, only scheduled public transport vehicles are involved. WLAN and UMTS networks covers the whole region, while VANET covers partially, only within a specified range of bus stops. Additionally, vehicle-to-vehicle communication is not considered;*

Assumption 2. *For any of the network selections, there is a predefined bandwidth and range-defined network selection is only possible if those conditions are met. Every user has different preferences based on their requirements. We will use the AHP method to assess each user's requirements and preferences.*

4.1. Case Study I

We will consider Auckland Central as shown in Figure 11 as an area for data analysis and as the locations to show the vehicle's distribution among different bus stops. We have considered four different locations: City Center, Britomart, Wellesley Street, and Auckland Hospital. All of these bus stops are equipped with local storage for data storage to upload or download onto buses on that route. Furthermore, all users' profiles are checked, as per the source and destination location of the data transmission, and buses are selected based upon that.

We will evaluate the performance of the AHP method using simulations over MATLAB, based upon different utility values for all attributes. We simulate for our goal to have optimal network selection based upon different criteria and alternatives. User preferences play an important role in the selection of the best available network in a heterogeneous environment. The proposed method for determining the user's preference is based upon the basic idea of AHP.

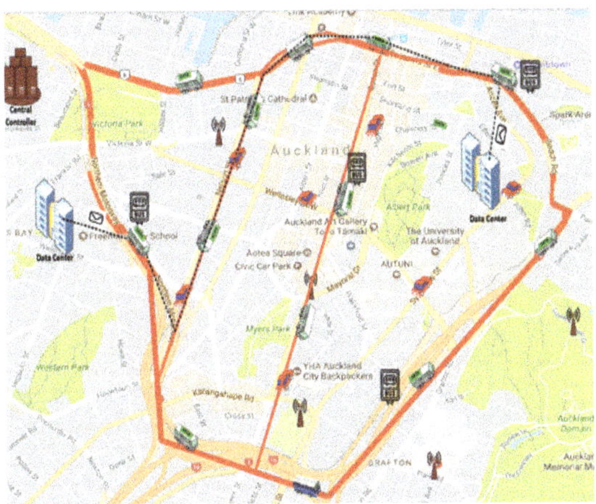

Figure 11. Auckland Central map with locations.

4.1.1. Service 1

We have categorized our services with all the different criteria as defined below. The first service is to be initialized from the data center (DC1) to the data center (DC2). The controller helps to select the optimal network as per their preferences being decided for different criteria defined above, such as Energy Efficiency (EE), Bandwidth (B), Delivery probability (DV), and Delay Tolerance (DT). The utility values have been defined for all attributes in Section 2, above.

$$S_1 = < EE, BDV, DT > \tag{20}$$

In this Service 1 (S_1), we assume that this service is for non-real-time applications, such as video surveillance data, that are accumulated at the data center and that can be delayed for up to 13 hours. There are three possible networks to choose from: UMTS, WLAN, and public transport. When the end-user sends and receives big data files, such as backup storage or large datasets, in TB or PB, this scheme applies. The scheme is, thus, more or less delay-tolerant, for example, background downloading of email messages, sending of data with Google Grive, and data backup. In this case, we give more importance to the energy-efficiency factor than other attributes, as we can bear delays for these applications or services. Now, the same procedure will be followed for all the attributes as per services, as defined in Tables 3–5.

Table 3. Pairwise Comparison Utility matrix as per importance scale.

Attributes	Energy Efficiency	Bandwidth	Delay Tolerance	Delivery Probability
Energy Efficiency	1	7	9	3
Bandwidth	1/3	1	7	2
Delay Tolerance	1/9	1/7	1	1/5
Delivery probability	1/3	1/2	5	1

Table 4. Normalized score table for all the attributes with the weight factor.

Attributes	Energy Efficiency	Bandwidth	Delay Tolerance	Delivery Probability	Critera Weight
Energy Efficiency	1	7	9	3	0.530345069
Bandwidth	1/3	1	7	2	0.164911216
Delay Tolerance	1/9	1/7	1	1/5	0.041457905
Delivery probability	1/7	1/2	5	1	0.280751063

Table 5. Normalized score table with priority vector.

Attributes	Energy Efficiency	Bandwidth	Delay Tolerance	Delivery Probability	Critera Weight	Priority Vector (P_w)
Energy Efficiency	1	3	9	7	0.530345069	0.5289
Bandwidth	1/3	1	7	2	0.164911216	0.1582
Delay Tolerance	1/9	1/7	1	1/5	0.041457905	0.0366
Delivery probability	1/7	1/2	5	1	0.280751063	0.2763

λ_{max} = 4.178069312; CI = 0.059356437; CR = 0.065951597 < 0.1

This pairwise matrix also passes a consistency check, which means that priority is selected correctly.

4.1.2. Service 2

The next service is more for the urgent delivery of data. In this case, the delay-tolerant indicator is about 3 hours, and the data volume is 64TB. As before, Service 2 (S_2) has similar attributes but different tendencies. This service includes real-time applications, such as Video-on-Demand. These services are delay-sensitive and, therefore, cannot be delayed for more than 3 hours. However, due to the large volume of data, we still grant more importance to energy efficiency and delay attributes than other attributes. It is the service class with the highest QoS requirements, and it switches from one network to another quickly as per users' profiles, such as telephony speech, VoIP, video conferencing, and other real-time activities. If a user is connected to WLAN and loses connection, they can then switch to UMTS for QoS. The same procedure will be followed for all attributes for Service 2, as defined in Tables 6–8.

Table 6. Pairwise Comparison Utility matrix as per importance scale.

Attributes	Energy Efficiency	Bandwidth	Delay Tolerance	Delivery Probability
Energy Efficiency	1	7	1	5
Bandwidth	1/7	1	1/7	2
Delay Tolerance	1	7	1	7
Delivery probability	1/5	1/2	1/7	1

Table 7. Normalized score table for all the attributes with the weight factor.

Attributes	Energy Efficiency	Bandwidth	Delay Tolerance	Delivery Probability	Criteria Weight
Energy Efficiency	1	7	1	5	0.42274576
Bandwidth	1/7	1	1/7	2	0.08567345
Delay Tolerance	1	7	1	7	0.45678945
Delivery probability	1/5	1/2	1/7	1	0.06435676

Table 8. Normalized score table with priority vector.

Attributes	Energy Efficiency	Bandwidth	Delay Tolerance	Delivery Probability	Critera Weight	Priority Vector (P_w)
Energy Efficiency	1	7	1	5	0.42274576	0.4163
Bandwidth	1/7	1	1/7	2	0.08567345	0.0782
Delay Tolerance	1	7	1	7	0.45678945	0.4455
Delivery probability	1/5	1/2	1/7	1	0.06435676	0.0599

λ_{max} = 4.156390957; CI = 0.052130319; CR = 0.057922576 < 0.1.

This pairwise matrix also passes a consistency check, which means that priority is selected correctly. We have calculated the weight for all three types of services by users' preferences for different attributes.

4.1.3. Service 3

The next service is different from the previous two. In this case, the delay tolerance is 6 hours and the data volume is 32TB. Service 3(S_3) has consistent attributes but different characteristics. This service is not that low in data volume, compared to the others. These services are delay-sensitive and, therefore, cannot be delayed for more than 6 hours. In this case, the user has all three options to disseminate data. The controller will first look for all the network options, including WLAN, UMTS, and whether there are buses available to carry data within the given timeframe. The same procedure will be followed for all attributes, as defined in Tables 9–11:

Table 9. Pairwise Comparison Utility matrix as per importance scale.

Attributes	Energy Efficiency	Bandwidth	Delay Tolerance	Delivery Probability
Energy Efficiency	1	1/6	1/6	1/7
Bandwidth	6	1	3	1
Delay Tolerance	6	1/3	1	1/5
Delivery probability	7	1	5	1

Table 10. Normalized score table for all the attributes with the weight factor.

Attributes	Energy Efficiency	Bandwidth	Delay Tolerance	Delivery Probability	Criteria Weight
Energy Efficiency	1	1/6	1/6	1/7	0.05355183
Bandwidth	6	1	3	1	0.36439882
Delay Tolerance	6	1/3	1	1/5	0.15369319
Delivery probability	7	1	5	1	0.4540202

Table 11. Normalized score table with priority vector

Attributes	Energy Efficiency	Bandwidth	Delay Tolerance	Delivery Probability	Criteria Weight	Priority Vector (P_w)
Energy Efficiency	1	1/6	1/6	1/7	0.05355183	0.0459
Bandwidth	6	1	3	1	0.36439882	0.3613
Delay Tolerance	6	1/3	1	1/5	0.15369319	0.1499
Delivery probability	7	1	5	1	0.45402002	0.4429

λ_{max} = 4.234869383; CI = 0.078289794; CR = 0.08698866 < 0.1.

This pairwise matrix also passes a consistency check, which means that priority is selected correctly. We have given importance to different attributes as per different services. Next, we calculated criteria weight for all the attributes and then, added priority vector to all of the attributes as per different services. Based on these calculations, next, we will rank our network for different services.

4.2. Network Selection for Different Services

We have discussed the AHP procedure and utility theory for all the attributes' weighings and preferences. Now, the AHP procedure will help us weigh different attributes for all of our services. In our work, we define the traditional and vehicular networks as alternatives to choose from and the available list is $I_{an} = (W, U, V)$. Algorithm 1 illustrates the whole process for optimal network selection based upon different services.

There is a list of available networks $I_{an} = (W, U, V)$ to choose from. We collect all the network attributes in list $I_{an} = a_1, a_2 \ldots . a_n$, named energy-efficient e_u, delivery probability dp_u, delay demand d_u, and available bandwidth b_i of both networks. Then, we follow all the steps to rank the network among all the networks, as per different services. We use this network-selection technique only to offer the best option as per their requirements to maintain QoS. It is mandatory to pass the consistency check in AHP in order to obtain an accurate judgment matrix. If any of the matrices fail to pass this check, the user will have to give preferences to the design matrix. We will first analyze public vehicle distributions near bus stops to know the availability of networks to choose from, and then further evaluate the performance of all networks for different services. Figure 12 illustrates the criteria weights given to all the attributes as per different services. For example, as discussed before, Service 1 has delay-tolerant features and will be considered an energy-efficient data-dissemination network. Therefore, the criteria weight will be allocated more heavily on the EE attribute. In such a way, all the weights are distributed as per the service profile. The priority vector is calculated for all the services as per different attributes as shown in Figure 13. The final score is calculated as discussed in Figure 14. Utility functions are defined already for all the attributes. For all of these services, we will have different utility values. We will score our network based upon the maximum utility value for all the services.

Algorithm 1: Optimal Network Selection

Input : Different services as per user's profile: energy efficient e_u, delivery probability dp_u, delay demand d_u, available bandwidth b_i of both networks, available network list I_{an}.

Output: Decision factor weight and rank of selected newtork, energy efficient weight w^e, bandwidth weight w^b, delivery probability weight w^{dp}, delay weight w^d.

1 According to the different services of users, build the decide hierarchy structure $P = x_{(1,j)}, x_{(2,j)}, ..., x_{(M,j)}$;

2 Loop 1: Construct decision Matrix P;

3 Loop 2: Calculate the weight of hierarchy $x_{(M,j)}$; including energy-efficient weight w^e, bandwidth weight w^b, data volume weight $w^d v$, and the delay weight w^d of the heirarichy;

4 Decide whether hierarchy $z_{(i,j)}$ is consistent;

5 If not, go back to **Loop 1**;

6 If $z_{(i,j)} < N$, go back to **Loop 2**;

7 Calculate the total weights, then attain the energy-efficient weight w^e, bandwidth weight w^b, delivery probability weight w^{dp}, and delay weight w^d;

8 Decide whether the whole hierarchy is consistent; if not, go back to **Loop 1**;

9 Obtain the final priority vector for all attributes;

10 Rank the network-selection score;

11 Exit the procedure.

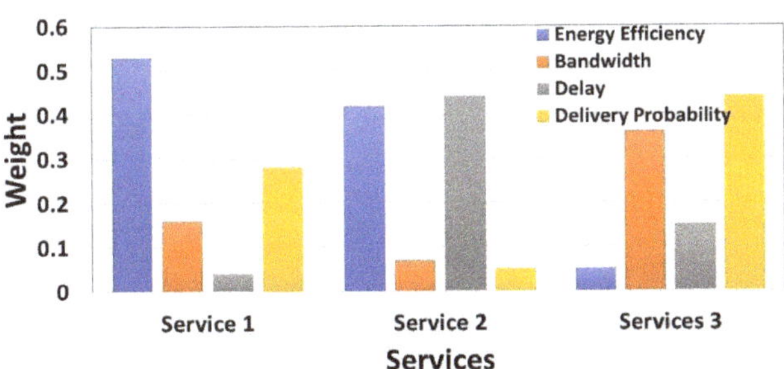

Figure 12. Weight distributed to all attributes as per different services.

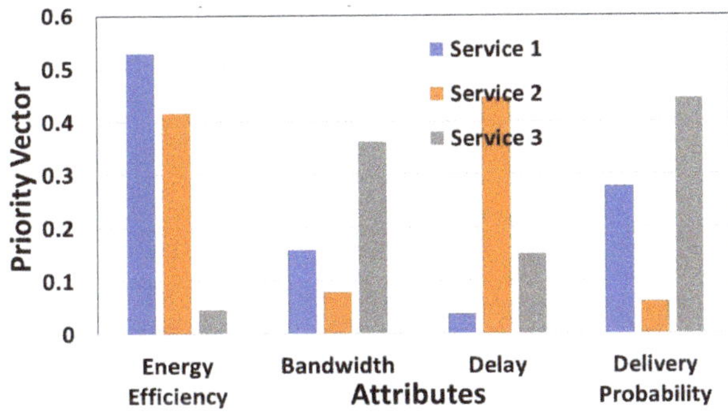

Figure 13. Priority vector for all services.

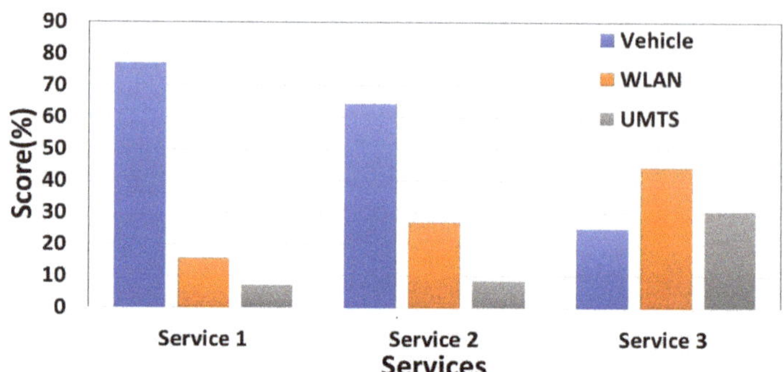

Figure 14. Network ranking for all services.

The AHP score is the final ranking of all the services as per the preferences given for all the attributes. For service 1, the ranking order is $Vehicle > WLAN > UTMS$, as the data is delay-tolerant and can be carried by vehicles for energy-efficient data-dissemination, as shown in Figure 14. However, service 2, which is delay-sensitive but for a larger volume than service 1, also gives preference to vehicular networks for data-delivery, rather than WLAN and UMTS, with a ranking order $Vehicle > WLAN > UTMS$. For service 3, the network ranking preference is in the order of $WLAN > UMTS > Vehicle$ for the urgent delivery of data so as to sustain QoS. In network dynamics, the most important factor is packet delivery without loss. Our heterogeneous network architecture guarantees the delivery of data by using any of the available networks and considering different attributes.

4.3. Case Study II

We conduct a numerical example to allocate different demands generated from DC on buses to carry data until the destination, while minimizing energy consumption. We consider 16 demands generated randomly from different bus stops to deliver their data carried by bus.

As shown in Figure 15, there are many bus stops around and demands have been allocated to the DC for data allocation onto the suitable bus. The controller will make an energy-efficient decision based upon Equation 13. DC is the central depot, where the bus begins and finishes its journey. As per Table 12, different data demands are generated for data being delivered from the parent stop to the destination stop. The controller identifies 4 buses, B1, B2, B3, and B4, to fulfill all demands with an energy-efficient solution. The total capacity of each bus is 150TB. The distance to each bus stop has been given from the central depot or source bus stop.

The demands must not exceed the maximum capacity of the bus. We use CVRP instances from the past and solve using the Cplex optimization solver. The Capacitated Vehicle Routing Problem is an NP-hard problem that can be solved exactly only for small instances. We have tested our objective function and observed an optimal solution while minimizing energy consumption. We assume that buses are available to carry data towards each bus stop. However, we will be calculating energy consumption while sending data through a traditional network to show via comparison that PTDD is an energy-efficient solution for delay-tolerant data applications.

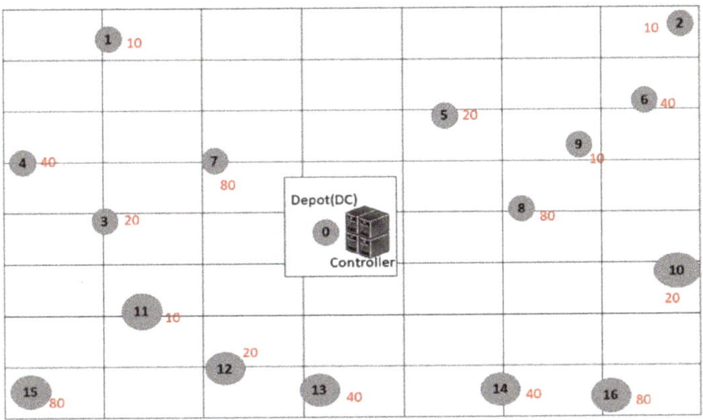

Figure 15. Proposed scheme for the problem.

Table 12. Demands from all bus stops for data allocation

Number of Buses per Day	Demands from Destination Stop (TB)	Distance from Depot (0) (Km)	Bus Capacity (TB)
1	10	5.48	150
2	10	7.76	150
3	20	6.95	150
4	40	5.82	150
5	20	2.74	150
6	40	5.02	150
7	80	1.94	150
8	80	3.08	150
9	10	1.94	150
10	20	5.36	150
11	10	5.02	150
12	20	3.88	150
13	40	3.54	150
14	40	4.68	150
15	80	7.76	150
16	80	6.62	150

As per the defined parameters, we have allocated data onto four buses that fulfill all requirements while minimizing energy consumption and returning to the source bus stop or depot after finishing their trips. We have defined all the bus routes with the optimal selected route for data allocation in Figure 16. All the buses have a maximum 150 TB capacity to carry and allocate data to all bus stops.

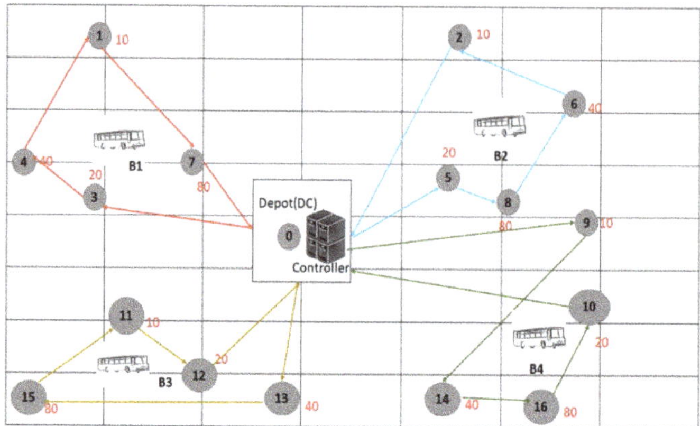

Figure 16. Data allocation onto each bus stop through buses.

Table 13 shows the computation results of all the buses traversing all the bus stops in a unidirectional format and the total distance covered during each trip.

Table 13. Set of test trips with the number of bus stops.

Bus Number	Selected Route	Total Distance Covered During the Trip
B1	0-3-4-1-7-0	12 km
B2	0-5-8-6-2-0	13 km
B3	0-13-15-11-12-0	12 km
B4	0-9-14-16-10-0	13 km

In our analysis, we have used 16 stops, which will be covered by four buses, to fulfill their demands being allocated from a DC to deliver data. In Figures 17 and 18, we can see that it is possible to disseminate data either from the core traditional network or PTDD in the heterogeneous network. However, if we have delay-tolerant data and can utilize public transport, PTDD is an energy-efficient solution. Bus stops 1 and 2 have demands of 10 TB, bus stops 3 and 5 have demands of 20 TB, bus stops 4 and 6 have demands of 40 TB, and bus stops 7 and 8 have demands of 80 TB.

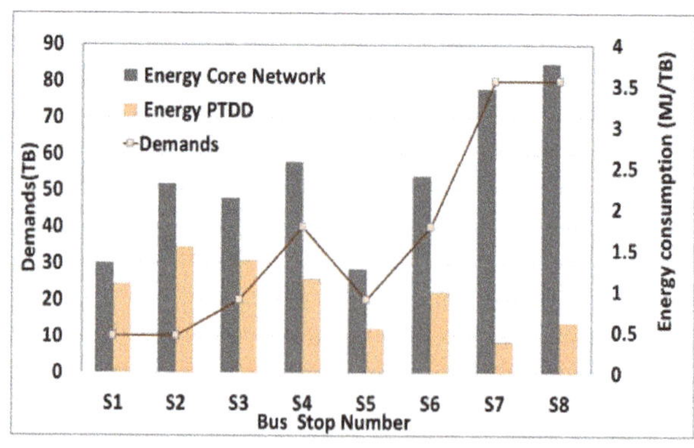

Figure 17. Energy consumption vs. bus stop number for generated demands.

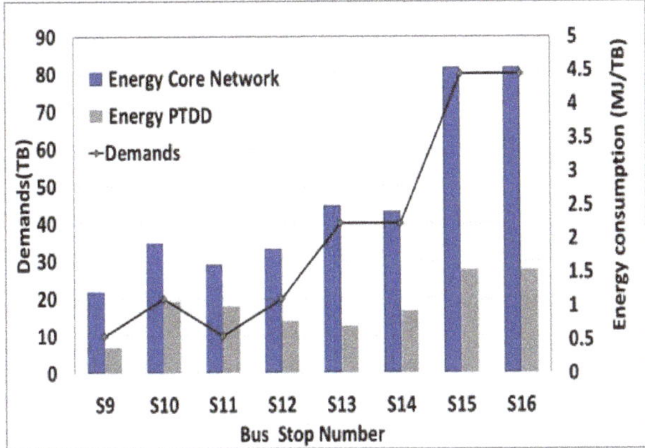

Figure 18. Energy consumption vs. bus stop number for generated demands.

As shown in Figure 18, bus stops 9 and 10 have demands for 10TB data, bus stops 10 and 12 demand 20 TB data, bus stops 13 and 14 demand 40TB data, and bus stops 15 and 16 have demands for 80TB data. The bus will carry and deliver data at each bus stop as per their demands. For the maximum demands of 80TB, we can analyze that the core network consumes 33% more energy than PTDD for data transmission.

A bus that stops for 500 seconds, for a total 60 buses, can offload 64.8 GB/day, with an effective throughput of 22.03 MB/s. Moreover, transmission performance is also highly influenced by the number of buses in a day and the stoppage time at a bus stop. Figure 19 shows the transmission performance of each network for different data rates. We have considered that public transport will be using IEEE 802.11ac as a network interface for data allocation. However, for comparison, we use the bandwidth of 512 MB/s and 1 GB/s in the traditional network to have a real difference. The outcome demonstrates that our proposed public transport network outperforms the traditional core network.

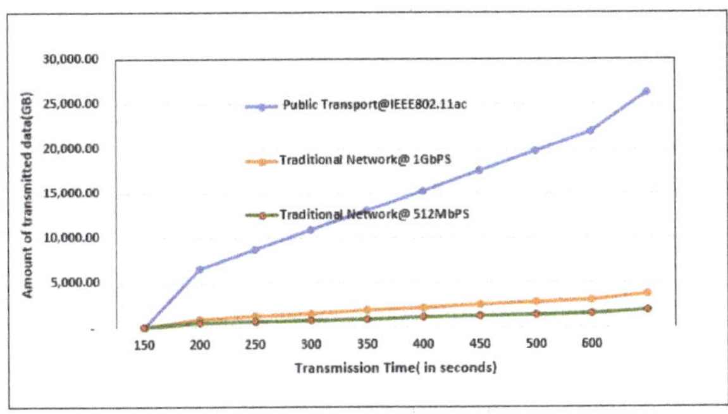

Figure 19. Transmission performance.

5. Discussion

We analyzed the various perspectives of traditional networks and every network has their standpoint and mode of communication. These networks rely mostly on big data analytics in the design of data communication networks. Therefore, these big data applications' survival would not be possible without the underlying support of networking, due to their extremely large volume and computing complexity. To elaborate further, we represent the three digital laws, Kryder's law, Moore's law, and Neilson's law, in Figure 20, which states that new products come into the market with each passing year with new technology. The basic idea of Kryder's law is to double the storage capacity every 12 months. Moore's law is somewhat like Kryder, but works on the processing speed of chips, which is doubled every 18 months. Moving forward to Neilson's law, which estimates that bandwidth doubles every twenty-one months, this last component of digital experience lags for both storage and processing speed. These three laws clearly explain that whatever new network technology comes onto the market, the available data (in online storage) is never fully accessed by the new network technology and the end-users. There will always be a gap between the available bandwidth and the available data/information storage online. This big data need will never be satisfied with internet technology.

This biggest challenge encourages the search for more connectivity options. Several attempts [41,42] have been made in developing efficient, sustainable, and integrated (wired/wireless) networks. The opportunistic network is one of the techniques to overcome this problem while disseminating data in-store and in a forward manner by connecting mobile devices. Many researchers have already discussed the concept of vehicular networks used as data carriers, as is discussed in the literature. However, we are contributing to existing work by introducing an alternative communication PTDD for sustainable data-dissemination via the introduction of a third layer of the public transport network to complement the conventional wired and wireless networks. For delay-tolerant data needs, our approach aims to better utilize the existing smart public vehicles and their parking spots with local storage to offload and upload data, thereby lowering the energy consumption while successfully delivering data.

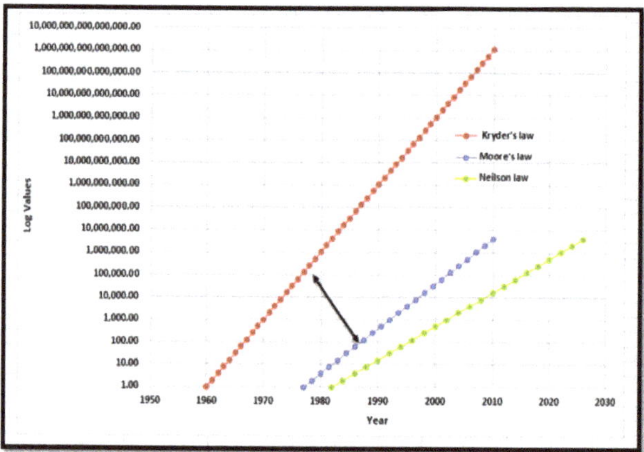

Figure 20. Kryder's, Moore's, and Nielsen's laws.

However, our work combines all of the networks, such as wired, wireless, and public transport, to use and switch according to the requirements of different services. The performance of our architecture was evaluated in two stages. First is the network selection among different networks, and second is when public transport is selected; in this stage, data is allocated onto these buses as per their fixed route. We evaluated our results using the SAS optimization tool while sending data using both networks and minimizing energy cost.

Our main, fundamental questions are: under which conditions would public transport will be selected among other networks? Relatedly, we consider how data will be allocated onto these buses. The existing literature has used many methods and compared them to show differentiation and their respective selection methods. In our work, the main implication is that the utility values are defined for all the attributes for the user's satisfaction, along with the AHP method for networks ranking. There is a vast amount of literature on existing networkselection techniques among different networks; we have utilized their concepts for public transport network selection based upon different user's demands for energy-efficiency. If we talk about the practical implications of our proposed system, any unexpected disaster—either naturally occurring or caused by human actions—firstly results in damage to the communication medium, although many of the technologies have been introduced for disaster management and attempt to connect the affected area with the rest of the world. However, in the post-disaster scenario, compared to the building and any bus stop, vehicles can be quickly moved to the affected area. In particular, public transport is firstly available to fulfill people's basic needs. Therefore, our PTDD can be efficiently utilized as a mobile communication backbone in disaster management.

6. Conclusions and Future Work

In this paper, we have presented an alternative communication channel PTDD for sustainable data-dissemination via the introduction of public transport networks to complement conventional wired and wireless networks. For delay-tolerant data needs, our approach aims to better utilize the PTDD and their parking spots/bus stops with local storage to offload and upload data. The controller used the MADM method to make an optimal network-selection decision among different networks and based on different services. The main implication is that the utility values are defined for all the attributes for the user's satisfaction, along with the AHP method for network ranking. We used Auckland's public transport network to prove that buses/public vehicles can be used as a data carrier. The results presented show the network ranking trends among all networks for different kinds of services. The second case study was presented using CVRP, which helped to minimize energy consumption with a fixed capacity of buses to allocate data onto each bus stop. This work provides strong evidence that significant energy savings can be achieved while still guaranteeing data delivery. The results presented here appear to be reasonable and promising, which ultimately proves that public transport can be used as another alternative communication network for delay-tolerant data needs. However, the proposed method could be affected by the highly dynamic changes in network topologies.

We have analysed PTDD with a static dataset; for future work, the network should be developed with dynamic factors such as traffic, weather, passenger flow data, etc., for real-time changes in the network. An analytical model for dynamic behaviors of bus movement would be a good future contribution. In terms of the future potential of applications, our system can be used in video surveillance systems. The transport agency has deployed people with cameras to record drivers illegally going into T3/T2 lanes. The public transport belongs to the same transport agency; thus, if these cameras can be deployed onto bus stops, then these buses can be utilized for carrying that accumulated data to the main center. These videos are not urgent and can be delayed up to hours for delivery. Hence, PTDD can be utilized efficiently to alleviate this network congestion case and to improve energy efficiency. However, the privacy and security part is lacking in our proposed work, and we will consider those aspects as future work and an extension of our proposed architecture.

Author Contributions: R.M. modeled the multi-attribute decision making for energy-efficient network selection in a heterogeneous network and adopted CVRP for optimal vehicle selection, implemented the case study, and analyzed the data under the supervision of W.L., X.L., J.G., and P.H.J.C. The manuscript was drafted by R.M., and was revised and proofread by W.L., X.L., J.G., and P.H.J.C. All authors have read and agreed to the published version of the manuscript.

Funding: This research received no external funding.

Data Availability Statement: Not Applicable, the study does not report any data.

Conflicts of Interest: The authors declare no conflict of interest.

References

1. Dehghani-Sanij, A.R.; Al-Haq, A.; Bastian, J.; Luehr, G.; Nathwani, J.; Dusseault, M.B.; Leonenko, Y. Assessment of current developments and future prospects of wind energy in Canada. *Sustain. Energy Technol. Assess.* **2022**, *50*, 101819. [CrossRef]
2. Babiceanu, R.F.; Seker, R. Big Data and virtualization for manufacturing cyber-physical systems: A survey of the current status and future outlook. *Comput. Ind.* **2016**, *81*, 128–137. [CrossRef]

3. Munjal, R.; Liu, W.; Li, X.J.; Gutierrez, J.; Chong, P.H.J. Telco asks transp: Can you give me a ride in the era of big data? In Proceedings of the 2017 IEEE Conference on Computer Communications Workshops (INFOCOM WKSHPS), Atlanta, GA, USA, 1–4 May 2017; IEEE: Piscataway, NJ, USA, 2017; pp. 766–771.
4. Deebak, B.D. Cooperative Mobile Traffic Offloading in Mobile Edge Computing for 5G HetNet IoT Applications In *Real-Time Intelligence for Heterogeneous Networks*; Springer: Berlin/Heidelberg, Germany, 2021; pp. 43–58.
5. Han, C.; Harrold, T.; Armour, S.; Krikidis, I.; Videv, S.; Grant, P.M.; Haas, H.; Thompson, J.S.; Ku, I.; Wang, C.X. Green radio: Radio techniques to enable energy-efficient wireless networks. *IEEE Commun. Mag.* **2011**, *49*, 46–54. [CrossRef]
6. Wang, X.; Vasilakos, A.V.; Chen, M.; Liu, Y.; Kwon, T.T. A survey of green mobile networks: Opportunities and challenges. *Mob. Netw. Appl.* **2012**, *17*, 4–20. [CrossRef]
7. Kelly, T.; Head, S. ICTs and Climate Change. In *ITU-T Technology, Tech. Rep*; 2007. Available online: https://www.itu.int/ITU-D/cyb/events/2008/geneva/docs/kelly-icts_and_climate_change-may2008.pdf (accessed on 22 December 2021).
8. Chen, Y. Great project overview,(slides) In Proceedings of the GreenTouch Open Forum (2011), Seoul, Korea, 8 April 2011.
9. Fettweis, G.; Zimmermann, E. ICT energy consumption-trends and challenges. In Proceedings of the 11th International Symposium on Wireless Personal Multimedia Communications, Lapland, Finland, 8–11 September 2008; Volume 2, p. 6.
10. McGreehan, J. Climate change and natural resources: What contribution can wireless communications make? In *UK Green Wireless Communication—Future Trend and Technology*; Springer: Berlin/Heidelberg, Germany, 2009; Volume 25, pp. 1–7.
11. Etoh, M.; Ohya, T.; Nakayama, Y. Energy consumption issues on mobile network systems. In Proceedings of the 2008 International Symposium on Applications and the Internet, SAINT 2008, Turku, Finland, 28 July–1 August 2008; IEEE: Piscataway, NJ, USA, 2008; pp. 365–368.
12. Li, M.; Si, P.; Zhang, Y. Delay-tolerant data traffic to software-defined vehicular networks with mobile edge computing in smart city. *IEEE Trans. Veh. Technol.* **2018**, *67*, 9073–9086. [CrossRef]
13. Baron, B.; Spathis, P.; Rivano, H.; de Amorim, M.D.; Viniotis, Y.; Ammar, M.H. Centrally controlled mass data offloading using vehicular traffic. *IEEE Trans. Netw. Serv. Manag.* **2017**, *14*, 401–415. [CrossRef]
14. Baron, B.; Spathis, P.; Rivano, H.; de Amorim, M.D. Vehicles as big data carriers: Road map space reduction and efficient data assignment. In Proceedings of the 2014 IEEE 80th Vehicular Technology Conference (VTC2014-Fall), Vancouver, BC, Canada, 14–17 September 2014; p. 1-5.
15. Wang, T.; Li, P.; Wang, X.; Wang, Y.; Guo, T.; Cao, Y. A comprehensive survey on mobile data offloading in heterogeneous network. *Wirel. Netw.* **2019**, *25*, 573–584. [CrossRef]
16. Bendaoud, F. A modified-SAW for network selection in heterogeneous wireless networks. *ECTI Trans. Electr. Eng. Electron. Commun.* **2017**, *15*, 8–17.
17. Salih, Y.K.; See, O.H.; Ibrahim, R.W.; Yussof, S.; Iqbal, A. A user-centric game selection model based on user preferences for the selection of the best heterogeneous wireless network. *Ann. -Telecommun.-Ann. Télécommun.* **2015**, *70*, 239–248. [CrossRef]
18. Sgora, A.; Vergados, D.D.; Chatzimisios, P. An access network selection algorithm for heterogeneous wireless environments. In Proceedings of the IEEESymposium on Computers and Communications, Riccione, Italy, 22–25 June 2010; IEEE: Piscataway, NJ, USA, 2010; pp. 890–892.
19. Wang, L.; Kuo, G.S.G. Mathematical modeling for network selection in heterogeneous wireless networks—A tutorial. *IEEE Commun. Surv. Tutor.* **2012**, *15*, 271–292. [CrossRef]
20. Bi, T.; Yuan, Z.; Trestian, R.; Muntean, G.M. URAN: Utility-based reputation-oriented access network selection strategy for HetNets. In Proceedings of the 2015 IEEE International Symposium on Broadband Multimedia Systems and Broadcasting, Ghent, Belgium, 17–19 June 2015; IEEE: Piscataway, NJ, USA, 2015; pp. 1–6.
21. Çalhan, A.; Çeken, C. An optimum vertical handoff decision algorithm based on adaptive fuzzy logic and genetic algorithm. *Wirel. Pers. Commun.* **2012**, *64*, 647–664. [CrossRef]
22. Fu, S.; Li, J.; Li, R.; Ji, Y. A game theory based vertical handoff scheme for wireless heterogeneous networks. In Proceedings of the 2014 10th International Conference on Mobile Ad-hoc and Sensor Networks, Maui, HI, USA, 19-21 December 2014; IEEE: Piscataway, NJ, USA, 2014; pp. 220–227.
23. Abid, M.; Yahiya, T.A.; Pujolle, G. A utility-based handover decision scheme for heterogeneous wireless networks. In Proceedings of the 2012 IEEE Consumer Communications and Networking Conference (CCNC), Las Vegas, NV, USA, 14–17 January 2012; IEEE: Piscataway, NJ, USA, 2012; pp. 650–654.
24. Goyal, R.K.; Kaushal, S.; Sangaiah, A.K. The utility based non-linear fuzzy AHP optimization model for network selection in heterogeneous wireless networks. *Appl. Soft Comput.* **2018**, *67*, 800–811. [CrossRef]
25. Jiang, D.; Huo, L.; Lv, Z.; Song, H.; Qin, W. A joint multi-criteria utility-based network selection approach for vehicle-to-infrastructure networking. *IEEE Trans. Intell. Transp. Syst.* **2018**, *19*, 3305–3319. [CrossRef]
26. Poirot, V.; Ericson, M.; Nordberg, M.; Andersson, K. Energy efficient multi-connectivity algorithms for ultra-dense 5G networks. *Wirel. Netw.* **2020**, *26*, 2207–2222. [CrossRef]
27. Sede, M.; Li, X.; Li, D.; Wu, M.Y.; Li, M.; Shu, W. Routing in large-scale buses ad hoc networks. In Proceedings of the 2008 IEEE Wireless Communications and Networking Conference, Las Vegas, NV, USA, 31 March–3 April 2008; IEEE: Piscataway, NJ, USA, 2008; pp. 2711–2716.
28. Xian, Y.; Huang, C.T.; Cobb, J. Look-ahead routing and message scheduling in delay-tolerant networks. *Comput. Commun.* **2011**, *34*, 2184–2194. [CrossRef]

29. Komnios, I.; Tsapeli, F.; Gorinsky, S. Cost-effective multi-mode offloading with peer-assisted communications. *Ad Hoc Netw.* **2015**, *25*, 370–382. [CrossRef]
30. Kassar, M.; Kervella, B.; Pujolle, G. An overview of vertical handover decision strategies in heterogeneous wireless networks. *Comput. Commun.* **2008**, *31*, 2607–2620. [CrossRef]
31. Song, Q.; Jamalipour, A. A network selection mechanism for next generation networks. In Proceedings of the IEEE International Conference on Communications, ICC 2005, Seoul, Korea, 16–20 May 2005; IEEE: Piscataway, NJ, USA, 2005; Volume 2, pp. 1418–1422.
32. Liang, G.; Guo, X.; Sun, G.; Fang, J. A User-Oriented Intelligent Access Selection Algorithm in Heterogeneous Wireless Networks. *Comput. Intell. Neurosci.* **2020**, *2020*, 8828355. [CrossRef]
33. Yu, H.; Ma, Y.; Yu, J. Network selection algorithm for multiservice multimode terminals in heterogeneous wireless networks. *IEEE Access* **2019**, *7*, 46240–46260. [CrossRef]
34. Munjal, R.; Liu, W.; Li, X.J.; Gutierrez, J. A Neural Network-Based Sustainable Data Dissemination through Public Transportation for Smart Cities. *Sustainability* **2020**, *12*, 10327. [CrossRef]
35. Munjal, R.; Liu, W.; Li, X.J.; Gutierrez, J. Big data offloading using smart public vehicles with software defined connectivity. In Proceedings of the 2019 IEEE Intelligent Transportation Systems Conference (ITSC), Auckland, New Zealand, 27–30 October 2019; IEEE: Piscataway, NJ, USA, 2019; pp. 3361–3366.
36. Poulliat, C. Mobile Data Offloading via Urban Public Transportation Networks. Ph.D. Thesis, Institut de Recherche en Informatique de Toulouse, Toulouse, France, 2017.
37. Nguyen-Vuong, Q.T.; Agoulmine, N.; Cherkaoui, E.H.; Toni, L. Multicriteria optimization of access selection to improve the quality of experience in heterogeneous wireless access networks. *IEEE Trans. Veh. Technol.* **2012**, *62*, 1785–1800. [CrossRef]
38. Saaty, T.L. Decision making with the analytic hierarchy process. *Int. J. Serv. Sci.* **2008**, *1*, 83–98. [CrossRef]
39. Vanier, D.; Tesfamariam, S.; Sadiq, R.; Lounis, Z. Decision models to prioritize maintenance and renewal alternatives. In Proceedings of the Joint International Conference on Computing and Decision Making in Civil and Building Engineering, Montréal, QC, Canada, 14–16 June 2006; pp. 14–16.
40. 2021. Available online: www.idc.com/ap/smartcities (accessed on 22 December 2021).
41. Heinzelman, W.R.; Kulik, J.; Balakrishnan, H. Adaptive protocols for information dissemination in wireless sensor networks. In Proceedings of the 5th annual ACM/IEEE International Conference on Mobile Computing and Networking, Seattle, WA, USA; ACM: New York, NY, USA, 15–19 August 1999; pp. 174–185.
42. Boukerche, A.; Darehshoorzadeh, A. Opportunistic routing in wireless networks: Models, algorithms, and classifications. *ACM Comput. Surv. (CSUR)* **2015**, *47*, 22. [CrossRef]

future internet

Article

Exploring the Benefits of Combining DevOps and Agile

Fernando Almeida [1,*], Jorge Simões [1] and Sérgio Lopes [2]

1   Polytechnic Higher Institute of Gaya (ISPGAYA) and INESC TEC, 4400-103 Porto, Portugal;
    jsimoes@ispgaya.pt
2   Polytechnic Higher Institute of Gaya (ISPGAYA) and Distance Education and eLearning Laboratory-Open
    University (LE@D-UAb), 1269-001 Lisbon, Portugal; ssargo@ispgaya.pt
*   Correspondence: almd@fe.up.pt

Abstract: The combined adoption of Agile and DevOps enables organizations to cope with the increasing complexity of managing customer requirements and requests. It fosters the emergence of a more collaborative and Agile framework to replace the waterfall models applied to software development flow and the separation of development teams from operations. This study aims to explore the benefits of the combined adoption of both models. A qualitative methodology is adopted by including twelve case studies from international software engineering companies. Thematic analysis is employed in identifying the benefits of the combined adoption of both paradigms. The findings reveal the existence of twelve benefits, highlighting the automation of processes, improved communication between teams, and reduction in time to market through process integration and shorter software delivery cycles. Although they address different goals and challenges, the Agile and DevOps paradigms when properly combined and aligned can offer relevant benefits to organizations. The novelty of this study lies in the systematization of the benefits of the combined adoption of Agile and DevOps considering multiple perspectives of the software engineering business environment.

Keywords: software development process; operations; software engineering; information system development; team structure

Citation: Almeida, F.; Simões, J.; Lopes, S. Exploring the Benefits of Combining DevOps and Agile. *Future Internet* **2022**, *14*, 63. https://doi.org/10.3390/fi14020063

Academic Editor: Davide Tosi

Received: 12 January 2022
Accepted: 17 February 2022
Published: 19 February 2022

Publisher's Note: MDPI stays neutral with regard to jurisdictional claims in published maps and institutional affiliations.

1. Introduction

The software development process can be viewed as a set of tasks required to produce high-quality software. The literature shows that the quality of the software produced reflects the way the process was carried out [1–3]. Although several software development processes exist, generic activities common to all of them stand out, as Sommerville [4] highlights, such as software specification (e.g., requirements definition, software constraints), software development (e.g., software design and implementation), software validation (e.g., software must be validated to ensure that the implemented functionality conforms to what was specified), and software evolution (e.g., software evolves as per customer need). The software development process provides an interaction between users and software engineers, between users and technology, and between system engineers and technology. In this sense, software development is an interactive learning process, and the result is an embodiment of knowledge gathered, transformed, and organized as the process is conducted.

A software development methodology includes a set of activities that assist in the production of software. These activities result in a product that demonstrates how the development process was conducted. Agile methodologies arise from the need to overcome the difficulties and disadvantages of applying traditional methodologies in project management and implementation. The Agile methodology assumes short periods of time between each delivery to ensure early and continuous delivery of software susceptible to evaluation [5]. In [6] it is also recognized that software implementation according to this paradigm is interactive and incremental, enabling early confirmation of whether or not the delivered artifact meets the needs and making the respective corrections with low risk and cost.

The main social and human factors involved in the adoption of Agile methodologies are the impact on organizational culture, namely by the collaborative culture imposed on developers and the implications of being embedded in an Agile team [7,8]. Constant feedback to all team participants on the activities being carried out, and the commitment to the team's goals, are highlighted in Junker et al. [9] as key elements for a well-functioning Agile team. Feedback and collective awareness are essential as opposed to individualism and lack of communication. This view is also confirmed by Sweetman and Conboy [10] when they highlight that feedback loops are the essence of the empirical and complex processes found in software engineering that require continuous adaptation based on learning obtained daily. Furthermore, complex projects are very unpredictable and therefore need a process that incorporates unpredictability [11].

Inspired by the success of Agile methods, the DevOps (Development and Operations) movement emerged that aims to take this line of reasoning to a higher level. This movement comes to break the traditional culture where there was almost no interaction between teams and, as highlighted by Luz et al. [12], the goal is to create a culture of collaboration between development and operations teams that allows increasing the flow of completed work. In summary, it is intended to increase the frequency of deployments while increasing the stability and robustness of the production environment. Beyond a cultural change, the DevOps movement also focuses on the practices of automating the various activities required to deliver quality code into production, such as creating environments for testing, automating testing, configuring infrastructure, data migration, auditing, and security, among others [13,14].

In the literature, we can essentially find studies on DevOps that explore ways to align development teams with operations [15], the benefits that this methodology can bring to organizations [16], and the challenges that are posed [17]. However, there is a research gap in the characterization of the simultaneous adoption of Agile and DevOps in organizations. In this dimension, the number of available scientific studies is limited and they mostly present individual views of their implementation, which does not allow for a sufficiently comprehensive characterization of the benefits of their combined adoption. We acknowledge the study conducted by Hemon et al. [18], which characterizes the different phases of Agile to DevOps transition (e.g., Agile, ongoing integration, and constant delivery), while Melgar et al. [19] explore the benefits of the combined SCRUM-DevOps approach in terms of increasing speed during the deployment process and increasing the quality of software processes. In both studies, there is just one empirical case study, which makes it difficult to generalize the findings. In this sense, this study seeks to bridge this research gap and presents an analysis of the benefits that can be found by the combined approach of DevOps and Agile considering a comprehensive set of twelve case studies that are representative of practices of simultaneous adoption of both methodologies. This approach supported by multiple case studies avoids the individual limits of each company's vision and reduces the risk of bias, and allows comparing, grouping, and systematizing the main benefits of the combined adoption of both methodologies.

The rest of this manuscript is organized as follows: Initially, a theoretical contextualization of the DevOps model is performed and the similarities between DevOps and Agile are explored. Next, the methodology and associated methods adopted in the study are described. This is followed by the presentation of the results and discussion of their relevance to the perception of the benefits of the combined adoption of DevOps and Agile. Finally, the conclusions are enumerated. It is also in this last phase of the manuscript that the limitations of the study are addressed and some suggestions for future work are provided.

2. Literature Review

2.1. DevOps Concepts and Model

In 2009, Paul Hammond and John Allspaw presented the talk "10+ Deploys Per Day: Dev and Ops Cooperation at Flickr" [20]. They explained how the developers' teams (Dev) and operations teams (Ops) could contribute to more agile and scalable software development. Tight integration between Dev and Ops to safely achieve several software deployments (more than 10) in a single day was a disruptive idea regarding software development and its evolution. Later, Patrick Debois coined the term DevOps (Development and Operations) and created the DevOps Day event [21]. Although the DevOps movement has been discussed for more than a decade [13–15,22] it still lacks a unique formal definition. For Wiedemann et al. [23], the lack of a unique definition could be intentional to allow each team to choose the definition that better suits its needs. Nevertheless, several authors proposed definitions such as the one by Leite et al. [13]— "DevOps is a collaborative and multidisciplinary effort within an organization to automate continuous delivery of new software versions while guaranteeing their correctness and reliability"—and others view it as a combination of values, principles, methods, practices, and tools [24]. Some other common definitions can be found in [23].

One of the key points in the execution of a project is the approach used to manage it. The traditional approach based on the waterfall model looks at the project in a linear way with several events, while in the iterative approach the development of software is undertaken through successive progress [25]. Therefore, it is common that the system is presented still incomplete or with some deficient parts. The objective is that the refinement of the product happens in stages until the desired result is achieved. The software development process does not end with the release of the code, but only when it closes the feedback loop between those who write the code and those who use it. DevOps aims to remove the barriers between traditionally independent teams: development and operations. Under the DevOps approach, these teams should work together across the entire software life cycle, from development and testing through deployment to operations. More than only a technical subject, DevOps deals with the organizational and human issues that arise in the software life cycle. It promotes a culture of collaboration, integration, and communication between teams to reduce the disconnect between them while assuring the delivery of software in an agile, safe, and stable way. According to Rajapakse et al. [14], the DevOps movement is currently a widely adopted software development approach having as a major benefit the ability to deploy releases more frequently and at a higher rate.

Some related concepts used in conjunction with DevOps are Continuous Integration, Continuous Delivery, and Continuous Deployment. As noted in [14], these concepts are considered key practices within DevOps, but are not always clearly used, as stressed by Stahl et al. [24]. Continuous Integration is a development practice where developers frequently integrate the code they produce, that has successfully passed testing, to the project under development [24]. Those integrations occur typically once a day. Continuous Delivery is a development practice where the software is kept in a reliable deployable state at any time [14]. Potentially, after every change, the software can be released. This leads to several release candidates that are evaluated. The deployment to production is made manually by a team member, with the appropriate authority, who decides when and which candidate should be released [14]. Apart from Continuous Delivery, in Continuous Deployment, release candidates resulting from software changes are automatically deployed to production without the intervention of any team member [14,24].

Regarding how organizations could adopt DevOps and measure its success, the CALMS framework is considered a foundational model for DevOps. CALMS is an acronym for Culture-Automation-Lean-Measurement-Sharing. CALMS was created by Jez Humble, co-author of The DevOps Handbook. The acronym highlights the five core elements of CALMS [23]:

- Culture: a cultural change focusing on collaboration and integration between developers' team and operations' team;

- Automation: a high level of automation to achieve continuous delivery running each code change through automated tests;
- Lean: the application of lean principles to increase efficiency and reduce complexity;
- Measurement: keeping key performance indicators for measuring performance and identifying where improvements can be achieved;
- Sharing: knowledge and best practice should be shared in the organization and across organizational boundaries.

Security issues concerning DevOps led to the spread of another term: DevSecOps. It adds "Security" to "Development" and "Operations", which were already part of the DevOps term. According to Rajapakse et al. [14], security is often treated as a non-functional requirement, handled at a later stage of the software development life cycle. Under DevSecOps, security should be built into every part of the DevOps life cycle. The purpose of the DevSecOps philosophy is to align the speed of code delivery with building secure code, merged into one streamlined process.

The application of DevOps still must deal with some problems and concerns that can limit its use (e.g., resistance to change, organizational vision, legacy systems) [26]. Misuse of the term, lack of clear guidelines and, as already mentioned, the lack of a clear definition creates some ambiguity about how to use DevOps principles. Those principles presuppose that, before DevOps, development teams and operations teams were working independently with almost no knowledge about each other's work. This lack of knowledge across teams is not, in general, as deep as DevOps assume. The whole software development process is improved with better collaboration between teams, as DevOps advocates, but it does not mean that DevOps teams did not previously cooperate. Another concern around DevOps is that its adoption rate is still low.

There is a close relation between DevOps and Agile methods in software development. According to Leite et al. [13], DevOps is an evolution of the Agile movement since software development under Agile favors small release iterations with customer reviews.

2.2. Similarities and Differences between DevOps and Agile

In the context of Software Engineering, as discussed in this paper, DevOps can be understood as a behavioral evolution of Agile development [27], which was gradually conceived through practical experiences of implementation in software development. However, it is important to point out that the Agile method has its focus directed specifically to software development [28], while DevOps aims to involve the software development area in the implementation and operation of the software developed or still under development, which shows us that DevOps processes are being implemented within the Agile processes.

Currently, in the professional community of Information and Communication Technologies (ICT) there is a growing consensus, in practice, that DevOps can be understood as "DevOps = Agile + Lean + IT service management (ITSM)" [29]. In its method and processes, DevOps adopts characteristics of frameworks related to the technical area of Agile software development together with ICT management processes. Complementarily, other relevant methodologies (e.g., Extreme Programming, Dynamic Systems Development Method, Kanban, SCRUM) have approaches intrinsically related to the Agile philosophy [30–33]. SCRUM is very well-known and intensely used in the Agile method, and is generally enhanced by the Kanban tool, for managing the workflow of software development, but which also fits very well into the DevOps development process itself [34]. Table 1 presents the problems or gaps in the Agile method that are solved by adopting the DevOps method.

Table 1. Problems with Agile development and DevOps solution (adapted from [35]).

Problem with Agile Development	DevOps Solution
Delivery of new features to the customer is often delayed.	DevOps tools are used to test and release new features as they are completed.
Completed software components are not compatible with each other.	Open interfaces and test automation make it possible to divide development into independent yet compatible parts.
Quality of the product is not ensured properly prior to release.	DevOps tools and practices help automating quality assurance and reduce the need for repetitive manual work.
New features break old functions.	The quality of existing functions is ensured quickly and automatically after each change.
Budget goals and deadlines are missed.	The tools and procedures of DevOps increase the transparency and predictability of the development work.
Developer teams and IT operations crews are not cooperating.	Developer teams and IT operations crews agree upon responsibilities together. Their goals are unified.

In the context of the Agile method workflow, under the SCRUM framework implementation, DevOps presents four metrics [36] that are directly related to software delivery through enhanced software engineering: i—waiting time, ii—deployment frequency, iii—mean time to restore (MTTR), and iv—change failure percentage. These metrics can be implemented in the execution cycle of the Agile by SCRUM approach, enhancing the qualitative process of the organizational performance of software companies, in meeting the planned objectives, because it increases in a relevant way the level of monitoring and maturity of the activities performed in each work cycle.

Organizations that implement Agile methodologies focus on productivity and process optimization by reducing execution time [37]. However, software development environments are shrouded in constant change, with continuous interactions of teams that focus on deliverable products. DevOps processes can improve the interactivity of the development teams, improving the integration process of the stakeholders of a project in an Agile environment, facilitating the continuity of the processes in a more balanced and stable way.

According to [38], the DevOps culture can be implemented to carry out an incident management process for deliverable products, allowing Agile development teams to be continuously monitored, because it integrates the software deployment process by monitoring operations, as provided by the DevOps framework. Therefore, DevOps enhances the stability of the Agile cycle, as we mentioned before; that is, while Agile focuses on productivity in product deliverables in a more technical approach around ICT [39], DevOps directs its processes to checking the level of effectiveness of what is produced in the work cycles of the development teams. Therefore, in this central aspect of integration, there will be a tendency toward an improved qualitative increment in the scope of Software Engineering.

Within what we have discussed so far, it was verified that the processes originated from Agile methodologies are the structural base of DevOps [18,39], and that the union of these methodologies increases the level of intelligence of the information system that is generated in the work cycles, because it implements several functions, such as Developers, System Architects, Product Owners, Release Engineers, and Testers. Therefore, the level of collaboration is elevated with professionals that are beyond the initially foreseen roles, for example, in the SCRUM approach that is currently used in Agile, promoting the creation of cross-functional teams in the context of the DevOps approach.

The approach of development teams that implement Agile methodologies, to operations teams as proposed by DevOps, tends to accelerate the software release process, with studies [40] indicating that in addition to acceleration, there is an increase in software quality in terms of reliability and maintainability. As a result, the deliverable product meets the conditions foreseen in the essential objectives of the project, from software development

(Agile), as in deployment and testing in operations (DevOps). However, despite DevOps decreasing the gap between Developers and Operators, in terms of standardization, the DevOps methodology lacks a simple approach or a roadmap to be followed for its implementation in an organization [41], leaving it up to companies to define their standards and metrics. This can present as a complicating factor, very dependent on the maturity level of organizations and work teams, which will have to define their specific integration processes in a DevOps approach.

3. Methodology

This study adopts a qualitative methodology to explore the benefits of the combined adoption of DevOps and Agile. This type of methodology is used in the context of social sciences and engineering and, according to Merriam and Tisdell [42], is concerned with a level of reality that cannot be quantified, exploring meanings, aspirations, beliefs, values, and attitudes that correspond to a deeper space of relationships, processes, and phenomena, and cannot be reduced for operationalization of variables. Dyba et al. [43] recognize that the software industry presents lines of research that are not only limited to exploring technical software engineering issues, but also need to look at the challenges of the intersection between technical and non-technical aspects. In this sense, adopting a purely quantitative approach is insufficient. Moreover, phenomena addressed in the field of project development in a DevOps and Agile paradigm are complex and unique. Therefore, the qualitative approach adopted in this study allows for exploring and understanding the relationships and activities performed by organizations in their software development activity.

In the scope of this study, twelve case studies were incorporated that report the simultaneous inclusion of DevOps and Agile methodologies in their software development teams. These cases correspond to reports and press releases from commercial vendors of reference in this area. The data come from secondary sources and the authors have not made any changes to the press releases that represent the view of each manufacturer. No summarization of the press releases has been made. It cannot be guaranteed that there is no risk of bias since the identified benefits come from press releases from commercial companies in the area and that have commercial goals in the market. However, to minimize this risk, external and internal validity mechanisms were used. To ensure external validity, multiple case studies were used of companies in different geographic areas, and to ensure internal validity, the same identifier was used to associate similar benefits between the case studies.

Table 2 presents brief descriptions of the profiles of the case studies. In general, it can be concluded that the area of activity of the organizations is similar through the offer of IT products and services based on cloud architecture. We also highlight the offer of consulting services in the field of web and mobile development, and the offer of complementary services in the field of artificial intelligence, big data, and cybersecurity.

Thematic analysis is an interpretive data analysis method widely used in the social sciences and engineering and was adopted in the context of this study to identify common benefits in the combined adoption of DevOps and Agile. Flexibility is, according to Braun and Clarke [44], one of the benefits of thematic analysis. Through its theoretical freedom, the thematic analysis provides a flexible and useful research tool that can potentially provide a rich and detailed set of information about the data. Figure 1 shows the process of conducting data. The step begins with the coding process in which one coded the advantages present in each case study report. After that, an iterative process follows in which common themes of the advantages among the case studies are identified. Finally, the last step consists of accounting for the benefits identified in the case studies. This approach allows for a ranking of the top benefits.

Table 2. Profiles of the case studies.

ID	Country	Description
CS1	India	A company that operates in the IT outsourcing services market in building technological solutions in areas such as web apps, mobile apps, cloud strategy, analytics and business intelligence, testing, quality assurance services, and Agile project management. Their report looks to different ways to enable DevOps in Agile environments.
CS2	USA	Information technology company and advanced training for IT professionals in the fields of programming and technological development. Their report explores the relevance of Agile principles for deployment activities.
CS3	Canada	A global company that offers consulting services to help companies adjust their development teams by integrating new practices and technologies. Their report explores how Agile practices should be updated considering the needs of operations teams in organizations.
CS4	USA	Cloud services provider offering technology infrastructures based on public cloud, private cloud, hybrid cloud, and multi-cloud. Their report explores the difference and similarities between both paradigms.
CS5	Australia	Australian software company that develops products for software developers, project managers, and other software development teams. Their press release presents how automation processes can be implemented using a combination of both paradigms.
CS6	USA	Global business and technology consulting firm dedicated to helping organizations leverage emerging technologies and the latest business management thinking to achieve competitive advantage and mission success. Provides consulting and training services, primarily targeted at executives. Their article explores the differences between the two paradigms and suggests points of convergence between them.
CS7	USA	Company specialized in the dissemination of technological information in the field of information and communication technologies. Their press release looks to important aspects observed while combining DevOps and Agile.
CS8	USA	Multinational company in information technologies that develops automation solutions and advanced knowledge in areas such as automation, enterprise DevOps, data-driven business, and adaptive cybersecurity. Their article explores the role that Agile practices can play in DevOps.
CS9	Switzerland	Company that operates mainly in the European market in providing captivating scalable cloud-based solutions. Their article looks at the isolated benefits of each paradigm and tries to predict the benefits of their combined adoption.
CS10	India	Company that develops technological solutions for the education field and relies on the application of the Agile scalability paradigm, especially the SAFe model. Their article explores the change-driven management approach and looks at how DevOps and SCRUM address this challenge.
CS11	Germany	IT company that operates in the global market implementing cloud solutions, DevOps, software testing, quality assurance, artificial intelligence, and big data. Their press release looks at the problems in software engineering that the joint adoption of both paradigms can solve.
CS12	UK	A consulting company that aims to optimize work processes in organizations using cloud solutions, slack, and Trello services. Their press release looks at the role of the cloud and Agile methodologies in developing the DevOps paradigm.

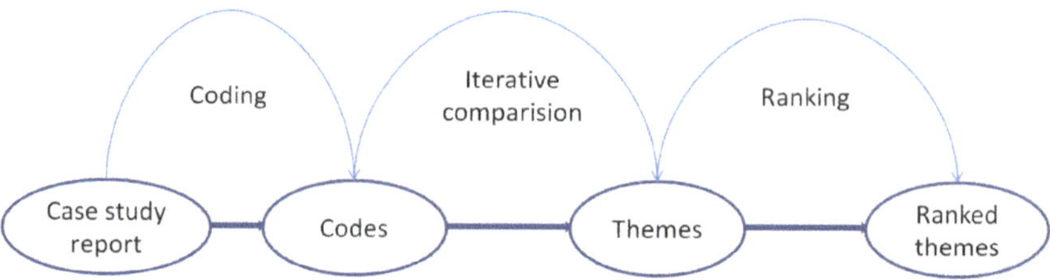

Figure 1. Thematic analysis process (authors own illustration).

4. Results

The twelve case study reports associated with each company presented in Table 2 were thoroughly read and each identified benefit was assigned a unique identifier. Each theme is identified by the acronym "BF" and a number is associated to differentiate each benefit. Common themes have the same acronym. A brief description of how each benefit is understood in the case studies is also included. Table 3 presents the identification process of the themes associated with each case study. Table 3 shows the themes for all the case studies mentioned in Table 2. We highlight the existence of themes that are common to several case studies. Although the themes are common, the vision of each case study in relation to them has some relevant oscillations, which indicate a complementary vision of the institutions present in the case studies. For example, in CS1 time to market emerges due to increased collaboration between teams, whereas in CS8 process integration is highlighted. Something similar emerges in relation to cost. Cost reduction is understood in CS6 as achieved from the existence of multi-skilled human resources, whereas in CS8 cost reduction is motivated by the effects of increased team performance.

Table 3. Identified benefits themes in the case studies.

ID	Benefit
CS1	(BF1) Time to market: greater collaboration between teams reduces software delivery cycles (BF2) Automation: the combined development and production process becomes more automated to meet market needs
CS2	(BF2) Automation: continuous delivery and integration combined with fast releases lead to the automation of activities (BF3) Communication: promote constant communication between development and operational team (BF4) Mindset and culture: establishment of collaboration among teams
CS3	(BF1) Time to market: through continuous delivery from the development phases (BF5) Planning: the product backlog now includes services are products that need to be deployed, scalable, maintained, monitored, and supported as a service
CS4	(BF4) Mindset and culture: increase the quality of collaboration (BF6) Visibility: more visibility for release and upgrade processes (BF7) Risk mitigation: better identification of risks in the context of each sprint (BF8) Software quality: decrease the existence of software errors and helps to launch faster patches
CS5	(BF2) Automation: contribution for the implementation of Agile fluency model which focus on value, transparency, and alignment (BF3) Communication: amplify feedback loops between development and operational team (BF4) Mindset and culture: looks to the performance of all system instead of local departments. Furthermore, promotes learning from failure. (BF5) Planning: increase the planning dimension of unplanned events typically found in the context of operational teams

Table 3. *Cont.*

ID	Benefit
CS6	(BF1) Time to market: deployment chains cut the time needed to get a product to market (BF9). Cost: combining people and activities makes people more multi-skilled with future reduction in human resource costs
CS7	(BF2) Automation: increasing code size and complexity encourages process automation (BF3) Communication: communication between both teams is constant with feedback loops (BF10): Software quality: functional and load tests are both considered (BF11). Efficiency: project management considers performance metrics that result from combined methods in both areas
CS8	(BF1) Time to market: integrated processes make order fulfillment faster (BF6) Visibility: increased visibility over data and processes (BF9) Cost: increased productivity and team performance
CS9	(BF2) Automation: increase speed and agility to attend continuous requirements changes (BF3) Communication: smooth communication between the team and the customers by continual iteration (BF12) Flexibility: agility in the face of continuous requests for revision becomes important to make the organization competitive
CS10	(BF2) Automation: implementation of a paradigm based in continuous integration, continuous delivery, and continuous deployment (BF3) Communication: by fostering communication in the teams, constant collaboration is promoted
CS11	(BF2): Automation: shorten the development cycle by promoting the automation of repetitive tasks (BF10) Software quality: focus on end-product quality
CS12	(BF2) Automation: better performance when compared against on-premise DevOps automation. Furthermore, it contributes to eliminates human errors (BF12) Flexibility: it empowers each stage of the application delivery lifecycle

Table 4 summarizes the comparative analysis of the identified benefits. All previously identified benefits are mapped. The "ranking" attribute allows us to understand the relative importance of the benefits and to perceive which ones are transversal to several case studies and which ones emerge only in a very specific context of each organization. The benefits related to automation (BF2), communication (BF3), and time to market (BF1) stand out. These are the three main benefits that can be found in the combined adoption of Agile and DevOps. Conversely, there are other benefits that are identified in a smaller number of case studies, namely, those related to efficiency (BF11), risk mitigation (BF7), and software quality (BF8). These benefits are less relevant and arise in the specific context of each organization, which indicates that they are more difficult to replicate in other software companies.

Table 4. Comparative analysis of benefits and ranking.

Benefit	CS1	CS2	CS3	CS4	CS5	CS6	CS7	CS8	CS9	CS10	CS11	CS12	Ranking
BF1	X		X			X		X					#3
BF2	X	X			X		X		X	X	X	X	#1
BF3		X			X		X		X	X			#2
BF4		X		X	X								#4
BF5			X		X								#5
BF6				X				X					#5
BF7				X									#10
BF8				X									#10
BF9						X		X					#5
BF10							X				X		#5
BF11							X						#10
BF12									X			X	#5

5. Discussion

Although Agile and DevOps are widespread and different concepts, they can be combined and offer relevant benefits to organizations. As reported in [45], companies have problems in the process of implementing and releasing new software versions because most of the time this is a process performed manually. In addition, this approach leads to a high quantity and frequency of errors [46]. To reduce the incidence of problems and increase flexibility and automation, non-operational resources can be used and in environments that are not in production. The combined adoption of Agile and DevOps allows the developer to gain greater control over the environment, infrastructure, and applications.

The seamless integration between Agile and DevOps generates a more collaborative and Agile framework. This approach leads to a simplification and automation of model processes to make them more rational and efficient. A classic example of this benefit is given by Fabro [47] when highlighting the reduction in delivery cycles, endowing small development packages with a previously unrecognized value. Hemon-Hildgen et al. [48] also highlight the role of orchestration, which consists of automating tasks to optimize the process and reduce repetitive steps that add little to the development cycle. Finally, automated testing along the Agile and DevOps chain allows the reuse of tests between environments and makes them more sustainable [49].

Team communication is recognized in DeFranco and Laplante [50] and Schmutz et al. [51] as the main cause for product delivery failures. By starting to work together, teams can more easily track the evolution of processes from their inception, which fosters the emergence of process improvements. Cois et al. [52] recognize that the great differentiator of DevOps lies in its ability to optimize communication between the teams involved and the customer. This allows, for example, the team to involve the operations team, which enables the implementation of the ITOps model [53]. This enables it to provide a sufficiently secure development environment. However, interconnecting it with an organization's Agile teams offers more potential. For example, the marketing and sales departments can be involved in the activities covering the delivery of the releases, which allows companies to add even more value to the product by using the full potential of their available resources.

The findings further revealed a very diverse number of benefits, such as increased visibility over processes, better identification and mitigation of risks, or increased software quality. The integration of the two paradigms fosters consolidation, which allows project managers to have greater visibility of both the work of the teams and the interdependencies between them [54]. Furthermore, iterative planning between teams makes it easier to adapt in case of changes, and continuous customer feedback generates value from the beginning of the project, lowering the risks associated with development and operation [55]. In the joint Agile and DevOps paradigm, both teams share responsibility for producing functional and quality code and need to work together to achieve these common goals.

Finally, it is recognized that in recent years there has been a growing adoption of the term "DevOps culture", as a counter position to DevOps implementations based only on tools. In DevOps culture, it is advocated that software development and infrastructure teams work together towards the same goal [56,57]. As Clavier and Kaminski [58] argue, DevOps not only optimizes development processes but changes the way employees think about their products and interact with customers. The combined Agile and DevOps approach allows the leveraging of these benefits by enhancing empathy among team members and unites sectors that previously worked independently and without personal connection. Furthermore, as Venugopal [59] acknowledges, when there is trust between teams, then it also increases the freedom that professionals have to experiment and innovate, without the problems of incompatibility and miscommunication as there would be with separate teams.

6. Conclusions

This study demonstrates that the Agile and DevOps paradigms are not incompatible but can bring benefits to organizations when properly aligned. While Agile brought about a fast delivery model aligned with customer expectations, DevOps optimized this system. In this sense, an alternative that usually gives great results is the adoption of both methodologies. They not only complement each other but also help companies to face changes in a team.

Changing the strategy and methodology of a team can be a delicate process full of obstacles. Therefore, organizations must address this challenge in a cross-cutting way within the organization to avoid isolated silos that do not contribute to collaborative work. Agile creates a space for more agile work with partial deliveries, while DevOps creates an environment conducive to managing these processes, with effective communication.

This study offers both theoretical and practical relevant contributions. In the theoretical dimension, this study has enabled the identification of a set of benefits of the combined adoption of both paradigms through the adoption of multiple case studies of software companies in the international market. The study identifies a total of 12 benefits and allows us to explore the relative relevance of each of them. From a practical perspective, the benefits identified are relevant to companies that, having adopted Agile and DevOps alone, have not yet taken steps towards the combined adoption of both models. The findings of the study made it evident that the two models are not incompatible, but when combined they can amplify their impacts on organizations.

Limitations and Future Research Directions

This study presents some limitations. Firstly, the case studies included in this study come from secondary sources, which does not allow us to deepen the knowledge on the themes with the use of interviews that may evidence the application of both paradigms. Furthermore, the case studies come from companies with commercial purposes, which may not give a totally unbiased view of the benefits to the organizations or represent very specific groups of the population. Nevertheless, this study adopted external and internal validation mechanisms to reduce this risk of bias. As future work it is recommended that the business view be complemented with a scientific view of the benefits of combining DevOps and Agile and, to this end, a systematic review of the literature in the field can be conducted. Moreover, the qualitative approach used does not allow us to systematically identify all the advantages and make a rigorous quantification of these benefits. As future work, a quantitative study based on a large dataset is suggested to identify the advantages of the combined adoption of both paradigms considering also different sectors of activity of the organizations. It would also be relevant to consider the degree of maturity in the implementation of Agile and DevOps in these organizations and, thus, explore its relevance in the benefits found, since it is expected that some of the benefits may be more easily achieved by organizations with lower levels of maturity in these processes.

Author Contributions: Conceptualization, F.A., J.S. and S.L.; methodology, F.A.; validation, F.A., J.S. and S.L.; formal analysis, F.A.; investigation, F.A., J.S. and S.L. writing—original draft preparation, F.A., J.S. and S.L.; writing—review and editing, J.S. and S.L. All authors have read and agreed to the published version of the manuscript.

Funding: This research received no external funding.

Institutional Review Board Statement: Not applicable.

Informed Consent Statement: Not applicable.

Data Availability Statement: Not applicable.

Conflicts of Interest: The authors declare no conflict of interest.

References

1. Dyba, T.; Dingsoyr, T. Empirical studies of agile software development: A systematic review. *Inf. Soft. Tech.* **2008**, *50*, 833–859. [CrossRef]
2. Ergasheva, S.; Kruglov, A. Software Development Life Cycle early phases and quality metrics: A Systematic Literature Review. *J. Physics. Conf. Ser.* **2020**, *1694*, 012007. [CrossRef]
3. Panwar, D.; Tomar, P.; Kumar, P. Innovative methods to make the component-based software development process more effective to produce quality software. *J. Stat. Manag. Syst.* **2017**, *20*, 765–775. [CrossRef]
4. Sommerville, I. *Software Engineering*; India Education Services: Bengaluru, India, 2018.
5. Shore, J.; Warden, S. *The Art of Agile Development*; O'Reilly Media: Newton, MA, USA, 2021.
6. Petersen, K.; Wohlin, C. A comparison of issues and advantages in agile and incremental development between state of the art and an industrial case. *J. Syst. Soft* **2009**, *82*, 1479–1490. [CrossRef]
7. Gregory, P.; Taylor, K. Defining Agile Culture: A Collaborative and Practitioner-Led Approach. In Proceedings of the IEEE/ACM 12th International Workshop on Cooperative and Human Aspects of Software Engineering (CHASE), Montreal, QC, Canada, 27 May 2019. [CrossRef]
8. Tolfo, C.; Wazlawick, R.S.; Ferreira, M.G.; Forcellini, F.A. Agile methods and organizational culture: Reflections about cultural levels. *J. Soft Maint. Evol. Res. Pract.* **2011**, *23*, 423–441. [CrossRef]
9. Junker, T.L.; Bakker, A.B.; Gorgievski, M.J.; Derks, D. Agile work practices and employee proactivity: A multilevel study. *Hum. Relat.* 2021; in press. [CrossRef]
10. Sweetman, R.; Conboy, K. Portfolios of Agile Projects: A Complex Adaptive Systems' Agent Perspective. *Proj. Manag. J.* **2018**, *49*, 18–38. [CrossRef]
11. Brink, T. Managing uncertainty for sustainability of complex projects. *Int. J. Manag. Proj. in Bus.* **2017**, *10*, 315–329. [CrossRef]
12. Luz, W.P.; Pinto, G.; Bonifácio, R. Building a collaborative culture: A grounded theory of well succeeded devops adoption in practice. In Proceedings of the 12th ACM/IEEE International Symposium on Empirical Software Engineering and Measurement, Oulu, Finland, 11–12 October 2018.
13. Leite, L.; Rocha, C.; Kon, F.; Milojicic, D.; Meirelles, P. A Survey of DevOps Concepts and Challenges. *ACM Comp. Surv.* **2019**, *52*, 127–162. [CrossRef]
14. Rajapakse, R.N.; Zahedi, M.; Babar, M.A.; Shen, H. Challenges and solutions when adopting DevSecOps: A systematic review. *Inf. Soft Tech.* **2022**, *141*, 106700. [CrossRef]
15. Wiedemann, A.; Wiesche, M.; Gewald, H.; Krcmar, H. Understanding how DevOps aligns development and operations: A tripartite model of intra-IT alignment. *Eur. J. Inf. Syst.* **2020**, *29*, 458–473. [CrossRef]
16. Jabbari, R.; bin Ali, N.; Petersen, K.; Tanveer, B. Towards a benefits dependency network for DevOps based on a systematic literature review. *J. Soft: Evol. Proc.* **2018**, *30*, e1957. [CrossRef]
17. Joby, P. Exploring DevOps: Challenges and Benefits. *J. Inf. Tech. Dig. World* **2019**, *1*, 27–37. [CrossRef]
18. Hemon, A.; Lyonnet, B.; Rowe, F.; Fitzgerald, B. From Agile to DevOps: Smart Skills and Collaborations. *Inf. Syst. Front.* **2020**, *22*, 927–945. [CrossRef]
19. Melgar, A.S.; Osores, J.; Osores, R.; Relaiza, H.R.; Flores, J.A.; Orihuela, V.H.; Lozano, R.A. DevOps as a culture of interaction and deployment in an insurance company. *Turk. J. Comp. Mat. Educ.* **2021**, *12*, 1701–1708. [CrossRef]
20. Hammond, P.; Allspaw, J. 10+ Deploys Per Day: Dev and Ops Cooperation at Flickr [Video]. 25 June 2009. Available online: https://www.youtube.com/watch?v=LdOe18KhtT4 (accessed on 28 December 2021).
21. Frederic, P. The Incredible True Story of How DevOps Got Its Name [Web Log Message]. 6 May 2014. Available online: https://newrelic.com/blog/nerd-life/devops-name (accessed on 28 December 2021).
22. Fitzpatrick, L.; Dillon, M. The Business Case for Devops: A Five-Year Retrospective. *Cutter. IT J.* **2011**, *24*, 19–27.
23. Wiedemann, A.; Forsgren, N.; Wiesche, M.; Gewald, H.; Krcmar, H. Research for Practice: The DevOps Phenomenon. *Com. ACM* **2019**, *62*, 44–49. [CrossRef]

24. Stahl, D.; Mårtensson, T.; Bosch, J. Continuous Practices and Devops: Beyond the Buzz, What Does it All Mean? In Proceedings of the 43rd Euromicro Conference on Software Engineering and Advanced Applications (SEAA), Vienna, Austria, 30–31 August 2017. [CrossRef]
25. Larman, C.; Basili, V.R. Iterative and Incremental Development: A Brief History. *Computer* **2003**, *36*, 47–56. [CrossRef]
26. Tozzi, C. 5 Problems with DevOps [Web Log Message]. 12 January 2021. Available online: https://www.itprotoday.com/devops-and-software-development/5-problems-devops (accessed on 28 December 2021).
27. Lwakatare, L.E.; Kuvaja, P.; Oivo, M. Relationship of DevOps to Agile, Lean and Continuous Deployment. In *Product-Focused Software Process Improvement*; Abrahamsson, P., Jedlitschka, A., Nguyen Duc, A., Felderer, M., Amasaki, S., Mikkonen, T., Eds.; Lecture Notes in Computer Science; Springer International Publishing: Cham, Switzerland, 2016; pp. 399–415. [CrossRef]
28. Wang, C.; Liu, C. Adopting DevOps in Agile: Challenges and Solutions. Adopting DevOps in Agile: Challenges and Solutions. 29 June 2018. Available online: http://www.diva-portal.org/smash/record.jsf?pid=diva2%3A1228684&dswid=5071 (accessed on 2 January 2022).
29. Galup, S.; Dattero, R.; Quan, J. What Do Agile, Lean, and ITIL Mean to DevOps? *Com. ACM* **2020**, *63*, 48–53. [CrossRef]
30. Hema, V.; Thota, S.; Kumar, S.N.; Padmaja, C.; Krishna, C.B.; Mahender, K. Scrum: An Effective Software Development Agile Tool. *IOP Conf. Ser. Mater. Sci. Eng.* **2020**, *981*, 022060. [CrossRef]
31. Santos, P.S.; Beltrão, A.C.; Souza, B.P.; Travassos, G.H. On the benefits and challenges of using kanban in software engineering: A structured synthesis study. *J. Soft Eng. Res. Dev.* **2018**, *6*, 13. [CrossRef]
32. Fojtik, R. Extreme Programming in development of specific software. *Procedia Comput. Sci.* **2011**, *3*, 1464–1468. [CrossRef]
33. Sani, A.; Arbain, A.F.; Jeong, S.R.; Ghani, I. A Review on Software Development Security Engineering using Dynamic System Method (DSDM). *Int. J. Comp. Applic* **2013**, *69*, 33–44. [CrossRef]
34. Mousaei, M.; Gandomani, T.J. DevOps Approach and Lean Thinking in Agile Software Development: Opportunities, Advantages, and Challenges. *J. Soft Eng. Int. Syst.* **2020**, *5*, 1–10.
35. Hamunen, J. Challenges in Adopting a Devops Approach to Software Development and Operations. 23 July 2016. Available online: https://aaltodoc.aalto.fi/handle/123456789/20766 (accessed on 2 January 2022).
36. Marnewick, C.; Langerman, J. DevOps and Organizational Performance: The Fallacy of Chasing Maturity. *IEEE Soft* **2021**, *38*, 48–55. [CrossRef]
37. Subramanian, A.; Krishnamachariar, P.K.; Gupta, M.; Sharman, R. Auditing an Agile Development Operations Ecosystem. In *Research Anthology on Agile Software, Software Development, and Testing*; International Management Association, Ed.; IGI Global: Hershey, PA, USA, 2022; pp. 1154–1176. [CrossRef]
38. Faustino, J.; Pereira, R.; Alturas, B.; Silva, M.M.D. Agile Information Technology Service Management with DevOps: An Incident Management Case Study. *Int. J. Agile Syst. Manag.* **2020**, *13*, 339–389. [CrossRef]
39. Dörnenburg, E. The Path to DevOps. *IEEE Soft* **2018**, *35*, 71–75. [CrossRef]
40. Céspedes, D.; Angeleri, P.; Melendez, K.; Dávila, A. Software Product Quality in DevOps Contexts: A Systematic Literature Review. In *Trends and Applications in Software Engineering*; Mejia, J., Muñoz, M., Rocha, Á., Calvo-Manzano, J.A., Eds.; Advances in Intelligent Systems and Computing; Springer International Publishing: Cham, Switzerland, 2020; pp. 51–64. [CrossRef]
41. Nybom, K.; Smeds, J.; Porres, I. On the Impact of Mixing Responsibilities Between Devs and Ops. In *Agile Processes, in Software Engineering, and Extreme Programming*; Sharp, H., Hall, T., Eds.; Lecture Notes in Business Information Processing; Springer International Publishing: Cham, Switzerland, 2016; pp. 131–143. [CrossRef]
42. Merriam, S.B.; Tisdell, E.J. *Qualitative Research: A Guide to Design and Implementation*; John Wiley & Sons: Hoboken, NJ, USA, 2015.
43. Dyba, T.; Prikladnicki, R.; Rönkkö, K.; Seaman, C.; Sillito, J. Qualitative research in software engineering. *Emp. Soft Eng.* **2011**, *16*, 425–429. [CrossRef]
44. Braun, V.; Clarke, V. *Thematic Analysis: A Practical Guide*; SAGE Publications: Thousand Oaks, CA, USA, 2021.
45. Danesh, A.S.; Saybani, M.R.; Danesh, S.Y. Software release management challenges in industry: An exploratory study. *Afri. J. Bus. Manag.* **2011**, *5*, 8050–8056. [CrossRef]
46. Ogheneovo, E. Software Dysfunction: Why Do Software Fail? *J. Comp. Commun.* **2014**, *2*, 25–35. [CrossRef]
47. Fabro, V. The Unified Value of Agile and DevOps. 14 December 2020. Available online: https://www.insight.com/en_US/content-and-resources/tech-journal/winter-2020/the-unified-value-of-agile-and-devops.html (accessed on 5 January 2022).
48. Hemon-Hildgen, A.; Rowe, F.; Monnier-Senicourt, L. Orchestrating automation and sharing in DevOps teams: A revelatory case of job satisfaction factors, risk and work conditions. *Eur. J. Inf. Syst.* **2020**, *29*, 474–499. [CrossRef]
49. Ali, N.; Daneth, H.; Hong, J.E. A hybrid DevOps process supporting software reuse: A pilot project. *J. Soft Evol. Proc.* **2020**, *32*, e2248. [CrossRef]
50. DeFranco, J.F.; Laplante, P.A. Review and Analysis of Software Development Team Communication Research. *IEEE Trans. Prof. Commun.* **2017**, *60*, 165–182. [CrossRef]
51. Schmutz, J.B.; Meier, L.L.; Manser, T. How effective is teamwork really? The relationship between teamwork and performance in healthcare teams: A systematic review and meta-analysis. *BMJ Open* **2019**, *9*, e028280. [CrossRef] [PubMed]
52. Cois, C.A.; Yankel, J.; Connell, A. Modern DevOps: Optimizing software development through effective system interactions. In Proceedings of the IEEE International Professional Communication Conference (IPCC), Pittsburgh, PA, USA, 13–15 October 2014. [CrossRef]

53. Kumar, N.; Gondkar, R. Role of ITOps in DevOps. In Proceedings of the International Conference on Innovative Computing & Communication (ICICC), New Delhi, India, 19–20 February 2021.
54. Reifer, D. Is Merging Agile and DevOps Worth the Pain? 17 January 2019. Available online: https://www.cutter.com/article/merging-agile-and-devops-worth-pain-501791 (accessed on 5 January 2022).
55. Ozanich, A. DevOps Lifecycle vs Agile Methodology: Learning the Difference. 18 November 2021. Available online: https://blog.hubspot.com/website/devops-vs-agile (accessed on 5 January 2022).
56. Ebert, C.; Gallardo, G.; Hermantes, J.; Serrano, N. DevOps. *IEEE Soft* **2016**, *33*, 94–100. [CrossRef]
57. Luz, W.P.; Pinto, G.; Bonifácio, R. Adopting DevOps in the real world: A theory, a model, and a case study. *J. Syst. Soft* **2019**, *157*, 110384. [CrossRef]
58. Clavier, P.; Kaminski, A. How We Applied a DevOps Mindset to Manage Our People Data. 15 January 2021. Available online: https://tdwi.org/articles/2021/01/15/biz-all-apply-devops-mindset-to-manage-people-data.aspx (accessed on 7 January 2022).
59. Venugopal, D. DevOps: Driving Innovation with Old Habits. 1 September 2020. Available online: https://devops.com/devops-driving-innovation-with-old-habits/ (accessed on 7 January 2022).

future internet

Article

Fast Library Recommendation in Software Dependency Graphs with Symmetric Partially Absorbing Random Walks

Emmanouil Krasanakis [1,*,†] and Andreas Symeonidis [1,2,†]

1   Central Macedonia, Aristotle University of Thessaloniki, 54124 Thessaloniki, Greece; symeonid@ece.auth.gr or asymeon@cyclopt.com

2   Cyclopt, Central Macedonia, 55535 Thessaloniki, Greece

*   Correspondence: manios.krasanakis@issel.ee.auth.gr

†   These authors contributed equally to this work.

Abstract: To help developers discover libraries suited to their software projects, automated approaches often start from already employed libraries and recommend more based on co-occurrence patterns in other projects. The most accurate project–library recommendation systems employ Graph Neural Networks (GNNs) that learn latent node representations for link prediction. However, GNNs need to be retrained when dependency graphs are updated, for example, to recommend libraries for new projects, and are thus unwieldy for scalable deployment. To avoid retraining, we propose that recommendations can instead be performed with graph filters; by analyzing dependency graph dynamics emulating human-driven library discovery, we identify low-pass filtering with memory as a promising direction and introduce a novel filter, called symmetric partially absorbing random walks, which infers rather than trains the parameters of filters with node-specific memory to guarantee low-pass filtering. Experiments on a dependency graph between Android projects and third-party libraries show that our approach makes recommendations with a quality and diversification loosely comparable to those state-of-the-art GNNs without computationally intensive retraining for new predictions.

Keywords: Software Library Recommendation; graph filters; dependency graphs; link prediction

Citation: Krasanakis, E.; Symeonidis, A. Fast Library Recommendation in Software Dependency Graphs with Symmetric Partially Absorbing Random Walks. *Future Internet* **2022**, *14*, 124. https://doi.org/10.3390/fi14050124

Academic Editor: Davide Tosi

Received: 3 April 2022
Accepted: 18 April 2022
Published: 20 April 2022

1. Introduction

The pervasive integration of mobile phones in everyday life and the digitization of practically all aspects of human activities have led to a constant need for new software services, applications and platforms. This need drives a highly motivated software development industry, whose aim is to cater quickly to user needs with new or repurposed software. In this regime, agile and component-based engineering practices are predominantly adopted [1] that reuse previously developed software and quickly share it between developers, mostly in the form of well-documented and tested libraries. These are distributed by online services such as the Maven repository of Java libraries [2], the PyPI repository of Python libraries [3], and the npm registry of Javascript libraries [4].

However, the sheer size of coding ecosystems/repositories [5] makes it a daunting prospect to find which libraries would best support new projects. For example, as of writing, Maven hosts more than three million software artifacts. In this setting, programmers, especially those working in unfamiliar domains, need to conduct time-consuming research through many libraries to select those suited to their needs, or else they risk incurring technical debt to their projects in the long run [6]. To reduce search effort, automated tools have been proposed to recommend which libraries to use (Section 2.1). This is often achieved by analyzing the dependencies between projects and libraries and adopting a collaborative filtering outlook [7] that recommends additional libraries based on those already used. For example, the inclusion of server-related libraries could imply potential interest in database management libraries frequently used together in other projects. Collaborative filtering approaches organize projects and libraries in graphs, whose edges correspond to project–library dependencies. These then mine structural patterns to recommend new dependencies, for example, with Graph Neural Networks (GNNs—Section 2.3).

As far as predictive accuracy is concerned, collaborative filtering approaches yield high-quality recommendations. To achieve this, they typically learn latent representations (e.g., embeddings) for all software projects and libraries and let pairwise project and library representation comparisons (e.g., the cosine similarity of their embeddings) rank libraries based on their similarity to projects under examination. The most similar libraries are considered to implement the functionality needed by projects and are thus recommended for adoption. However, when dependency graphs evolve with more libraries and—importantly—projects for which to make recommendations, these approaches need to be retrained to create representations accounting for new nodes and dependencies. In practice, this translates to usability costs by locking recommendation pipelines until training is over (Section 3).

To create deployment-friendly library recommendation services, in this work, we argue that collaborative filtering can be conducted with no-learning alternatives that make informed ad hoc assumptions about which co-usage patterns to mine. These alternatives sacrifice some predictive quality for the benefit of avoiding training and its associated costs. In particular, we look at graph filters (Section 2.2) to recommend libraries based on how structurally proximate they are within dependency graphs to libraries already being used. Graph filters are computationally efficient (their running times scale near-linearly with the number of dependencies and their outcomes can be quickly computed, even without high-end GPU hardware) and only rely on their chosen understanding of structural proximity. We specifically choose absorbing random walks' filters that emulate human-driven library discovery combining co-usage exploration and memory of previous discoveries (Section 4.3); we employ such filters with the goal of quickly finding libraries similar to those that humans use in their projects and hence reduce the effort of searching for these.

Our contribution lies in (a) proposing graph filters as a viable alternative to more sophisticated but ultimately unwieldy library recommendation tools; (b) analyzing which types of filters to employ for high-quality library recommendations; and (c) introducing a new variation of absorbing random walk filters, called symmetric partially absorbing random walks for link prediction that has no learnable parameters—not even hyperparameters. The usefulness of our approach is experimentally demonstrated on a large real-world dependency graph of third-party library dependencies, where it outperforms representation learning based on matrix factorization in terms of the predictive quality and diversification of results and lags only a little behind state-of-the-art GNNs that require computationally intensive retraining for every new recommendation task.

The rest of this paper is organized as follows. In Section 2, we overview the related literature and present theoretical concepts needed to position our analysis, namely from the domains of graph signal processing and GNNs. In Section 3, we showcase practical issues with deploying representation learning for library recommendation and explain how these can be resolved when switching to graph filters. Based on this explanation, in Section 4, we analyze real-world library discovery practices and selected the appropriate filters that are automated yet emulate human-driven discovery. To show the effectiveness of these filters, in Section 5, we organize experiments to compare our approach with existing alternatives. In Section 6, we discuss the experimental results in terms of real-world usefulness, address threats to validity, and point out promising future work. Finally, in Section 7, we summarize our work and conclude the paper.

2. Background and Related Work

2.1. Library Recommendation

Many recommendation system approaches are applied in the field of assisted software engineering [8,9]. Among other tasks, these have also been used to recommend relevant libraries to developers to commence their work. Originally, library recommendation tools were similar to other search engines in that they used query terms pertaining to project keywords (e.g., extracted from source code). However, state-of-the-art systems employ co-usage patterns of libraries to recommend new ones based on those already included in projects [10–12]. This is effectively a type of collaborative filtering [13] which eventually coalesced to the matrix factorization of the LibSeek tool [14].

In practice, library recommendation is conducted on project–library dependency graphs, where software projects and libraries are nodes that are linked based on usage. That is, projects are linked to libraries they import. In terms of collaborative filtering, which aims to produce item recommendations for users based on item co-usage patterns (e.g., being bought by the same users in e-commerce platforms), libraries would correspond to items and software projects to users. Assuming that only links between projects and libraries are captured and not dependencies between libraries, dependency graphs are bipartite and described by matrices $A_{bip} : P \times I$ of P rows and I columns, where P is the number of projects and I the number of libraries. Their elements obtain values $A_{bip}[u,v] = \{1$ if project u depends on library $v, 0$ otherwise$\}$. For this formulation, matrix factorization approaches aim to generate representation matrices $H_{proj} : P \times h$ and $H_{lib} : I \times h$ whose rows correspond to underlying h-dimensional representations (embeddings) of projects and libraries, respectively. These representations are trained so that the dot product (the dot product between representations also models cosine similarity if L2 normalization is applied on representations) is higher between projects and their used libraries than between projects and unrelated libraries. In matrix form, the representation matching would ideally be able to reconstruct the bipartite graph per:

$$A_{bip} \approx H_{proj} H_{lib}^{T} \tag{1}$$

For example, LibSeek learns to approximate this factorization through stochastic gradient descent [15] on a loss function that heuristically weighs the differences between matrix elements, introduces L2 regularization on the representation matrices, and penalizes dissimilar representations of libraries and projects of similar graph neighborhoods.

A natural evolution of matrix factorization is to detect more complex library co-usage patterns with Graph Neural Networks (GNNs—Section 2.3). This direction has only recently been explored with the introduction of GRec [16] and similar works that also account for metadata other than dependencies [17]. Approaches consider adjacency matrices $A : (P+I) \times (P+I)$ describing the bipartite dependency graphs per $A = [0_P A_{bip}; A_{bip}^T 0_I]$, where $0_X : X \times X$ are square matrices of zeros. Adjacency matrices are then input in the GNN link recommendation pipelines, such as the ones described in Section 2.3.

2.2. Graph Signal Processing

Graph signal processing [18] is a way to systematize information propagation in graphs through their edges. In particular, it starts from a similar definition of adjacency matrices as above $A = \{1$ if edge u, v exists, 0 otherwise$\}$, which is modified to be applicable to any type of graphs, not only bipartite ones. It then considers normalizations $\hat{A}$ that reduce the importance of profligately connected nodes' edges. One popular type of normalization is the symmetric expression

$$\hat{A} = D^{-1/2} A D^{-1/2} \tag{2}$$

where D are diagonal matrices of node degrees with elements $D[u,v] = \{\sum_{v'} A[u,v']$ if $u = v, 0$ otherwise$\}$. This regards edges (u,v) as bidirectional and re-weighs them by considering both endpoint degrees per $\hat{A}[u,v] = A[u,v]/\sqrt{D[u,u]D[v,v]}$.

Given adjacency matrix normalizations $\hat{A}$, graph signal processing explores information propagation through graphs by considering graph signals h_0 whose elements $h_0[u]$ hold values corresponding to nodes u. These values can be propagated to one-hop neighbors through the matrix-vector multiplication operation $\hat{A}h_0$. This is equivalent to the discrete signal processing shift operator (graph signal processing can model discrete signal processing if points in time are expressed as a line graph whose edges connect points with the next points) and is a type of additive aggregation across graph neighbor values, where neighbors v of nodes u are weighted by $\hat{A}[u,v]$. Iterating the shift operator k times per $\hat{A}^k h_0$ yields graph signal propagations k hops away from original values. Under this formalization, graph filters are defined as a weighted averaging of multi-hop propagation to obtain filtered signals h per:

$$h = F(\hat{A})h_0$$

$$F(\hat{A}) = \sum_{k=0}^{\infty} f_k \hat{A}^k \tag{3}$$

where $F(\cdot)$ is the graph filter and f_k are the weights placed on node values k hops away. Notably, symmetrically normalized adjacency matrices $\hat{A}$ can be decomposed into $\hat{A} = U\Lambda U^{-1}$, where Λ are diagonal matrices of eigenvalues $\lambda \in [-1,1]$ and U is the orthonormal base of eigenvectors. Applying graph filters on this decomposition yields:

$$F(\hat{A}) = \sum_{k=0}^{\infty} f_k (U\Lambda U^{-1})^k = \sum_{k=0}^{\infty} f_k U\Lambda^k U^{-1} = UF(\Lambda)U^{-1}$$

Hence, graph filters transform normalized adjacency matrix eigenvalues from λ to:

$$F(\lambda) = \sum_{k=0}^{\infty} f_k \lambda^k \tag{4}$$

Based on the above properties, spectral graph theory generalizes the concept of Fourier transformations to node-domain graph signals h_0 as $\mathcal{F}\{h_0\} = U^{-1}h_0$ and the inverse transform as $\mathcal{F}^{-1}\{h_0'\} = Uh_0'$. Analogously to traditional signal processing, graph (convolutional) filtering is defined as element-by-element multiplication $\odot$ in the Fourier domain. The node-domain equivalent of filtering can be written as convolution with a Fourier-domain filter $F(\bar{\lambda})$ as:

$$\mathcal{F}^{-1}\{F(\bar{\lambda}) \odot \mathcal{F}\{h_0\}\} = \mathcal{F}^{-1}\{F(\Lambda)\mathcal{F}\{h_0\}\} = UF(\Lambda)U^{-1}h_0 = F(\hat{A})h_0$$

where $\bar{\lambda}$ is the vector of the adjacency matrix eigenvalues and is considered its spectrum, whilst $F(\bar{\lambda})$ is applied on all spectrum dimensions.

Since Fourier-domain operations can be translated into node-domain filtering computations, graph filters are easy to implement [19] and require only an informed assumption of how the normalized adjacency matrix's spectrum needs to be transformed. For instance, two popular graph filters are (a) personalized PageRank [20–22], which arises from Markovian-like equivalents to random walks with restart within graphs and have parameters $f_k = (1-a)a^k$ controlled by one hyperparameter $a \in [0,1]$; and (b) HeatKernel [23,24] which emulates heat diffusion dynamics in graphs with parameters $f_k = e^{-t}t^k/k!$, where $k \in \{1,2,3,\dots\}$ is the number of hops away in which maximal importance is placed.

These filters are low-pass in the sense that parameters f_k are generally larger for smaller k, which in turn translates into a lesser impact on eigenvalues with absolute values closer to 0 than high-frequency eigenvalues with larger absolute values. In practical terms of node domain operations, low-pass filters place more emphasis onto diffusing node values of fewer hops away and thus introduce a type of graph signal smoothing that removes non-local implicit node relations, which can be thought of as high-frequency noise.

The above spectral analysis is tailored to the symmetric normalization of graph adjacency matrices and undirected graphs, i.e., for which $\hat{A}[u,v] = \hat{A}[v,u]$. However, some filters, such as personalized PageRank, are better known for non-symmetric normalizations arising from Markov chain modeling, such as $\hat{A} = AD^{-1}$, where graphs are defined by directed edges. Spectral theories are also available for these filters, but lay outside the scope of our work. Instead, when analyzing bipartite project–library graphs, we work with undirected edges that allow the transfer of graph signal values from both projects to libraries and libraries to projects (otherwise, filtering with graph signals would be stuck at recommending only immediate neighbors). Therefore, we adopt the adjacency matrix normalization of (2).

2.3. Graph Neural Networks for Link Prediction

Graph Neural Networks (GNNs) [25,26] are a popular machine learning paradigm that lets traditional feature-based neural network learning account for the relational information of data samples organized into graphs. This is achieved through message-passing protocols that gather and aggregate latent representations of graph neighbors, which are then transformed with neural network layers shared between all nodes before being passed on. Many industry-level applications focus exclusively on GNNs that employ the shift operation of graph signal processing as the aggregation operation, since the latter performs a (weighted) averaging of graph neighbor representations that can be efficiently implemented with sparse matrix multiplication within GPUs.

All GNNs start from initial matrices of node representations $H^{(0)}$, whose rows $H^{(0)}[u]$ correspond to features of nodes u. These could be unsupervised embeddings obtained by multilayer architectures and trained end-to-end [27] or other pre-processed machine learning features, such as weighted bag-of-word vectors. Then, given normalized adjacency matrices $\hat{A}$, convolutional GNNs average graph neighbor representations where these are weighted by corresponding edge weights. This kind of smoothing is understood as a natural extension of graph signal processing to vector-valued graph signals can be expressed in matrix form with the operation $\hat{A}H^{(\ell)}$, where $H^{(\ell)}$ are matrices of (latent) node representations.

Most GNNs add computational stability to the graph shift operation with a practice dubbed the renormalization trick. This adds self-loops to all nodes before computing the normalized adjacency matrix and will also be used throughout this work. Compared to the original matrix, the renormalization trick computes $\hat{A} = (I + D)^{-1/2}(I + A)(I + D)^{-1/2}$, where I the unit matrix. Since matrix multiplication can be efficiently computed by modern GPUs, especially if graphs are not fully connected and sparse representations can be leveraged to make computation time scales with the number of edges, convolutional GNNs have become a widely popular variety for analyzing the graphs of many nodes and edges [25,26].

Original GNN approaches (e.g., the architecture of Kipf and Welling [28] that helped popularize the domain) defined graph convolutional layers per:

$$H^{(\ell)} = \sigma\left(\hat{A}H^{(\ell-1)}W^{(\ell)}\right) \tag{5}$$

where $\sigma(\cdot)$ are nonlinear activation functions applied on matrices element-by-element, such as rectified linear unit activations $ReLU(x) = \max\{x, 0\}$ [29] and $W^{(\ell)}$ are learnable weights that help determine the output of GNN layers $\ell = 1, \ldots, L$. The output of the final layer is used for predictions, which for node classification have dimensions equal to the number of classes and arise from a softmax activation on the top layer to obtain an estimation of a binary one-hot encoding of class labels. On the other hand, for link prediction tasks, any number of latent representation dimensions can be outputted and compared pairwise to select the most similar pairs of nodes to recommend links for, for example, through a sigmoid activation of their dot product [30].

To avoid the oversmoothing of representations along multiple graph convolutions, state-of-the-art GNNs often include recurrent terms in the predictions. This is achieved either by adding feedback loops that trade-off between layer outputs (before being passed through the activation function) and $H^{(\ell)}$ with linear or feature-specific terms [31,32], or by aggregating the outcomes of all convolutional layers [27]. Recursive loops effectively inject the graph signal transformations of trained features in all layers.

One popular link prediction framework using GNNs is NGCF [27] which is also employed by the aforementioned GRec library recommendation system. This calculates the similarity between combined node representations found in the rows of the matrix:

$$H_{final} = H^{(0)} || H^{(1)} || \ldots || H^{(L)} \tag{6}$$

where $||$ represents the horizontal matrix concatenation and L the number of graph convolutional layers. That is, the elements of $(H_{final} H_{final}^T)[u, v]$ are considered the scores of linking nodes u and v (in the case of library recommendation, only project–library scores are kept from these to find the most related libraries to projects). The same framework also refines the convolutional layers of (5) with a self-attention mechanism to node layers per:

$$H^{(\ell)} = \sigma\big(\hat{A} H^{(\ell-1)} W^{(\ell)} + \hat{A} H^{(\ell-1)} \odot H^{(\ell-1)} W_{att}^{(\ell)}\big) \tag{7}$$

where $\odot$ represents the element-by-element matrix product with lesser priority than matrix multiplication and $W^{(\ell)}, W_{att}^{(\ell)}$ learnable parameters at layers ℓ.

3. Deploying Library Recommendation Services

As per all software services, it is important to look at the deployment and usage flows of library recommendation from a software engineering perspective. In this section, we analyze how well real-world systems can adopt the flows of (a) existing representation-based library recommendation algorithms overviewed in Section 2.1; and (b) no-learning algorithms that perform inference based on informed ad hoc assumptions. We introduced an algorithm of the second type in the next section. In both cases, we envisioned the deployment of algorithms as online (e.g., RESTful [33]) services that developers query to obtain recommendations for their software projects.

3.1. Deploying Representation Learning for Library Recommendations

Representation-based recommendation algorithms need to be retrained when new nodes are added to dependency graphs, so as to arrive at representations that implicitly capture both old and new node relational information. This does not scale well when many recommendation requests are made for services for new projects or project prototypes, for example, by many independent agile development teams. Accommodating requests for yet-unseen graph nodes (projects) is more important in software engineering compared to other domains where representation learning has been applied, because a primary use case is to aid the development of *new* software rather than altering existing projects. In fact, changing or integrating new dependencies mid-development requires rewriting software project components and is a form of technical debt.

Keeping the above in mind, let us consider the recommendation system flow of notifying users about interesting items, which is popular among previous library recommendation works, such as those overviewed in Section 2.1. These usually perform real-world evaluation by first creating recommendations on the whole corpus of software projects after one training run and then recommending those to developers. In practice, developer notifications about potentially useful libraries translate to service subscription models where developers sign up their projects and obtain periodic recommendations.

However, when deploying library recommendation systems "in the wild", subscription services neglect the practical needs of the software industry that require system interfaces to be queried at will and immediately produce results for new projects. This is particularly important for agile development, where delays to software project design, especially at the first exploratory or rapid prototyping stages, can undermine the whole development process [34]. At the same time, retraining accrues significant upkeep costs to keep being deployed, as representations need to be extracted periodically using computationally savvy hardware, such as GPUs or clusters of GPUs able to fit large dependency graphs in-memory. For example, if representation-based library recommendations were integrated in query-able online code repositories (e.g., GitHub), the latter would need to periodically retrain library and project representations on snapshots of dependency graph databases. Thus, to obtain recommendations for new projects, developers would need to first upload their implementations to be integrated in the databases and *wait* for the next training round to complete, as shown in Figure 1.

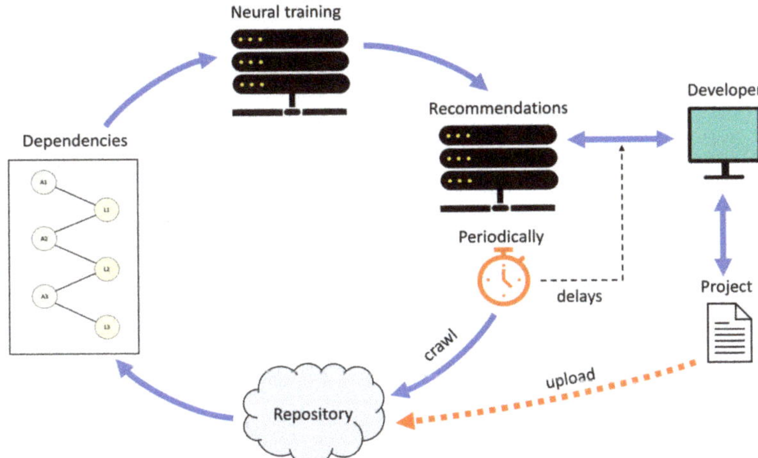

Figure 1. Integrating neural solutions in library recommendation pipelines. Periodic crawling of code repositories integrates uploaded projects in neural training, thus delaying developers from accessing recommendations for these projects.

To make matters worse, the above flow runs the risk of mining library usage from projects for which recommendation is the goal and existing dependencies are hastily selected. In particular, mining too many non-expert designs promotes co-usage pattern recommendations that replicate the perfunctory knowledge of early designs rather than well-maintained projects. To address this issue when designing real-world systems, there is an uncomfortable balance to be found between allowing any project as system input and letting hastily assembled projects (e.g., experimental versions during rapid prototyping) potentially ruin recommendation quality. Even in the best of cases, it is difficult to create tools that do not exclude the vast majority of experimental prototype queries. One realistic solution is for training to be conducted by integrating only a few low-quality projects in copies of dependency graphs, mining those for recommendations, and then discarding the integrated changes. However, this practice is unsustainable if library recommendation services are to become sufficiently popular for many hastily assembled recommendation requests to be made back-to-back; these would require a proportional number of training instances to run simultaneously.

Finally, beyond tangible workflow costs arising from long recommendation delays that are not able to immediately access recommendations could also discourage developers from adopting automated library recommendation. For instance, they could instead try to accelerate development cycles by ignoring automation and investing manual effort into library discovery instead. If so, the high usefulness of library recommendation systems—even high-quality ones— becomes obsolete once they fail to achieve high enough throughput.

3.2. Deploying No-Learning Library Recommendations

In this work, we propose moving away from representation learning and instead employing no-learning graph inference that only requires forward passes. Given that such algorithms exist and exhibit high enough predictive quality to be comparable to existing representation learning approaches, their recommendations can be computed on-demand for new libraries. For example, project and dependency metadata can be uploaded to no-learning recommendation systems to add them to dependency graphs just before inference takes place.

The advantage of no-learning algorithms is that, even if we consider the periodical mining of code repositories to extract new versions of dependency graphs, the latter do not make recommendation pipelines wait for their completion. In particular, for new project predictions, dependencies can be directly injected in the graphs before inference and removed afterwards—two operations with the minimal cost of, respectively, adding and removing one graph node and its edges. This is demonstrated in Figure 2, where the recommendation flow (the data flow cycle between the developer, the project, and the recommendation system) does not depend on the conclusion of periodical updates to recommend libraries.

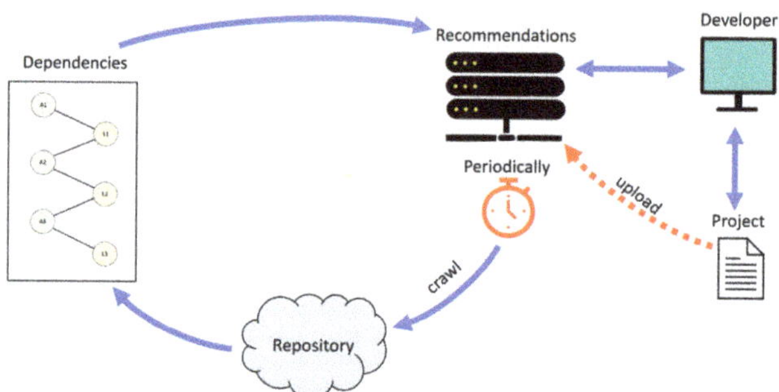

Figure 2. Recommendation pipelines based on no-learning graph inference. Periodic crawling only helps improve the quality of recommendations, and given that projects using similar libraries have already been crawled from code repositories, does not delay the recommendation.

Furthermore, the above-described recommendation flow can run in parallel to the mechanism extracting dependency graphs, such as by crawling repositories; if dependency graph snapshots already comprise enough usage patterns, mining the last known instead of the next graph would minimally affect recommendation outcomes given that the two differ only by a few nodes and edges. As a result, there would be a negligible impact on inference quality. By comparison, representation learning discussed in the previous subsection cannot make predictions with representations extracted by the last-known dependency graphs, because these do not have entries for new query projects.

Finally, given that repository crawling takes care to not extract dependencies from low-quality code (e.g., from recent projects with too few commits), the above flow sidesteps the issue of mining many confounding dependency patterns by undoing changes after inference. Since no training is required, and given that graph inference can be quickly computed, we envision that queue-based sequential pipelines can support high query loads before infrastructure parallelization (e.g., many servers providing access to the same recommendation service) is to be considered.

4. Graph Filters for Library Recommendations

In this section, we introduce graph filters as a collaborative filtering approach applicable to library recommendation. Although vanilla filters are often outperformed by state-of-the-art representation learning, we recognize that they also follow the no-learning paradigm described in the previous section. Thus, they fit well into the real-world sensibilities of deploying library recommendation services. Having identified this point, we look at the promising filters that were previously neglected by the recommendation system literature, but match high-level assumptions of how humans could go about mining project–library dependency graphs.

We start by describing the usage of graph filters for collaborative filtering and how these translate into our setting (Section 4.1). We then theorize which types of library co-usage patterns filters should model to emulate one potential human-driven library discovery process in hyperlink-like dependency graph exploration. In this regard, we identify the memory of past discoveries as a promising component often overlooked by previous approaches (Section 4.2). Finally, we translate our analysis to existing absorbing random walk filters, which model memory components for community detection tasks but have not been used in collaborative filtering, and infer node-wise memory strength to adhere to symmetric normalization principles instead of applying heuristics or training to determine it (Section 4.3). Experimental probing to demonstrate practical usefulness and to compare our approach to representation learning follows in the next section.

4.1. Revisiting Graph Filters for Collaborative Library Recommendations

We base our approach on collaborative filtering paradigms that run graph filters in bipartite graphs to find nodes relevant to ones of interest. Graph filters, especially personalized PageRank, were at some point a popular collaborative filtering tool [22], but this direction has in large part been abandoned in favor of the added accuracy offered by representation learning approaches such as GNNs. Other graph-mining tasks, however, have recently seen a resurgence of graph filters as equivalence to those is now understood as a primary contributor towards the efficacy of many GNN architectures [31,32,35,36]. Therefore, given that GNNs already boast a high predictive quality for library recommendations, we search for filters that are not lagging significantly behind with respect to predictive performance, while also satisfying our no-learning requirement.

To reconcile the opposite trends of collaborative filtering having abandoned graph filters and the latter being revisited by state-of-the-art research from other domains, we theorize that widely adopted graph filters are missing crucial assumptions that more sophisticated collaborative filtering mechanisms do not; these assumptions may not be as important in other predictive tasks, but are crucial for recommendation systems. In the next subsection, we identify lack of memory as one such assumption when emulating human-driven library recommendation.

We consider a general formulation of graph filters $F(\hat{A})$ that can take any functional form dependent on adjacency matrix normalizations $\hat{A}$ such as those described in Section 2. To recommend libraries for query projects with these, they need to parse project inputs. However, singleton data samples can lead to non-informed graph mining due to a lack of pairwise structural relations to mine. Thus, we employ the neighborhood inflation heuristic of Gleich et al. [37] to expand the search terms by including the immediate neighborhood of projects, i.e., their known dependent libraries, as query-able information to be included within graph signals. This kind of information was already sent to graph inference systems following the deployment of Figure 1 so that dependencies between the query project and at least one library are added to the dependency graph. Hence, there are no additional communication or computational costs associated with following this practice.

We hereby consider query graph signal h_0 with elements

$$h_0[v] = \{1 \text{ if } u = v \text{ or } v \text{ is a dependency of } u, 0 \text{ otherwise}\} \tag{8}$$

where u are the projects for which we provide library recommendations. Given these query signals, we pass them through filters of choice to obtain their structural proximity of all graph nodes $h = F(\hat{A})$. Finally, our methodology focuses on the proximity scores $h[v]$ of libraries v, where higher scores are structurally "closer" to target projects and thus indicate preferred recommendations.

4.2. Low-Pass Filters with Memory to Emulate Human-Driven Library Search

To design graph filters well-suited to library recommendation, we explore a search procedure within dependency graphs that emulates human exploration if no external sources of recommendation (e.g., expert guidance) was provided. In this, developers searching for the libraries best-fitting their projects look at projects using the same libraries and investigate which other dependencies are found there. This process is iterated to find projects and libraries that are more hops away, although presumably with lesser zeal, since after some time, irrelevant projects and libraries would start being found. To avoid getting "lost" in the dependency graph, the search would at some point restart. Up to this point, this process applies the popular random walk with restart search, whose probability of visiting nodes for stationary transition probabilities between pairs of nodes is proportional to the elements of graph signal outcomes of personalized PageRank [38].

We already discussed that recommendation systems based on personalized PageRank fail to reach a similar recommendation quality as more recent collaborative filtering approaches. For this reason, we argue that a missing assumption in the above exploration is the lack of memory during random walks. In particular, we propose that developers would not only backtrack during link-based exploration, but would also keep track of projects and libraries highly related to their query to also restart from there in future walks. Overall, we recognize four types of actions that can occur during human-driven random walks with restart and memory, given that developers would have arrived on a particular node: (a) visit a neighbor; (b) stay on the node; (c) remember the node; and (d) restart the random walk. These are visually illustrated in Figure 3.

Due to the chance of restarting random walks at all steps, it becomes progressively more likely to have restarted the more hops away developers move from query projects. In other words, recommendations will be more concentrated on libraries laying fewer hops away. In graph signal processing terms, the proposed filters would be low-pass and hence would not excessively smoothen the query across dependency graph edges and instead retain its original position within dependencies.

We stress that the theorization presented throughout this section is in large part derived by graph mining literature. Our contribution lies in identifying the key points best fitting the problem of automated library recommendation, and ultimately motivate the usage of appropriate graph filters in this setting.

4.3. Symmetric Absorbing Random Walks

In this subsection, we transcribe the above human-driven library discovery process to graph filters with minimal (ideally no) parameters. To do this, we make the assumption that all choices during random walks follow static distributions that only depend the nodes that developers are currently looking at. We also ignore real-world semantics, such as project names or descriptions, whose exploration is left for future work. Instead, we only use the structural characteristics of dependency graphs.

Given these assumptions, one possible tool to model random walks with memory are partially absorbing random walks [39]. Instead of only defining one type of filter, these introduce a framework for accounting for memory by letting a portion of random walks passing through nodes stay there. In terms of our theorization, this corresponds to developers remembering the nodes they visit and their relatedness to original queries. Various graph filters arise for different assumptions of how memory works, such as the probability of staying on nodes being proportional to node degrees, which is theoretically equivalent to personalized PageRank (more details below), or the alternative of assigning the same absorption rate to all nodes to retrieve tightly knit structural communities [39], where the absorption rate effectively describes the memorability of nodes.

Partially absorbing random walks account for the four types of random walk with memory actions described in the previous section and are recursively computable through the following formula:

$$h = S(S + \hat{D})^{-1}h_0 + S(S + \hat{D})^{-1}\hat{A}S^{-1}h \tag{9}$$

where $\hat{A}$ are symmetric normalizations of adjacency matrices presented in (2), S is a diagonal matrix whose diagonal elements $S[u, u]$ correspond to the absorption rates of nodes u, and $\hat{D}$ are diagonal matrices with elements $\hat{D}[u, v] = \{\sum_{v'} \hat{A}[u, v']$ if $u = v, 0$ otherwise$\}$.

We stress that $\hat{D}$ are the node degrees of the normalized (not the original) graph adjacency matrix. Correspondence between quantities appearing in (9) and the random walk procedure with memory in dependency graphs is demonstrated in Figure 3.

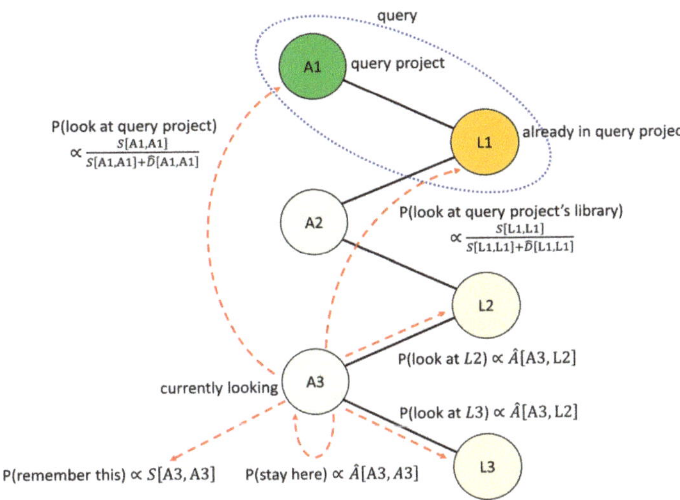

Figure 3. Random walks with memory within a project (A1,A2,A3)-library (L1,L2,L3) dependency graph. Dashed arrows represent the decisions available to developers when looking at project A3, given that they search for libraries for project A1 with known dependency L1.

Partially absorbing random walks can implement different graph filters, depending on chosen absorption rates S. For example, selecting $S = \frac{1-a}{a}\hat{D}$ for a parameter $a \in (0,1)$ reduces this scheme to the power method iteration of computing personalized PageRank, whereas $S = \frac{1-a}{a}I$ discovers tightly connected structural communities around query nodes with high probability [39]. In both cases, absorption rates only depend on one (hyper)parameter.

We now provide a novel way of selecting absorption rates. This starts by solving (9) with respect to h and expressing the graph signal outcome h of partially absorbing random walks per:

$$h = F(\hat{A})\hat{h}_0$$
$$F(\hat{A}) = \left(\mathcal{I} - S(S + \hat{D})^{-1}\hat{A}S^{-1}\right)^{-1}$$
$$\hat{h}_0 = S(S + \hat{D})^{-1}h_0$$

In this context, $\hat{h}_0$ is an adjusted version of the query graph signal that weighs the query project and its dependencies based on their absorption rates. $F(\hat{A})$ is the graph filter responsible for diffusing the adjusted query graph signal. Effectively, this can be expressed as a spectral filter $F(\hat{A}) = \hat{F}(\hat{\hat{A}}) = (\mathcal{I} - \hat{\hat{A}})^{-1}$, where $\hat{\hat{A}} = S(S + \hat{D})^{-1}\hat{A}S^{-1}$ is a new normalization applied on the normalized adjacency matrix $\hat{A}$ ($F(\hat{A})$ is not a spectral filter of $\hat{A}$ because it arise from a non-polynomial graph operations of the latter, but $\hat{F}(\hat{\hat{A}})$ is a spectral filter of $\hat{\hat{A}}$).

In the previous section, we formulated that graph signal filtering should be low-pass around query graph signals. To achieve this effect for $\hat{F}(\hat{\hat{A}})$, one simple solution would be to make $\hat{\hat{A}}$ symmetric. This way, and given that this matrix effectively has a non-negative shrunken version of $\hat{A}$'s elements, it would obtain eigenvalues λ in the range $\lambda \in [-1, 1]$, which in turn would be transformed into $\hat{F}(\lambda) = \sum_{k=0}^{\infty} \lambda^k$. Therefore, given that only positive absorption rates are accepted, i.e., visiting nodes lets developers retain at least *some* memory of them, a satisfactory condition to achieve a symmetric normalization $\hat{A}$ of the normalized adjacency matrix and hence a low-pass effect on the non-principal (i.e., those less than 1) eigenvalues of $\hat{A}$ can be computed per:

$$S(S + \hat{D})^{-1} = S^{-1} \Leftrightarrow S^2 - S - \hat{D} = 0 \Leftrightarrow S = \tfrac{1}{2}\left(\mathcal{I} + \sqrt{\mathcal{I} + 4\hat{D}}\right) \tag{10}$$

5. Experiments

In this section, we conduct the experiments to evaluate the efficacy of no-learning library recommendation compared to existing representation learning alternatives. We start by describing the evaluation dataset and measures used in experiments (Section 5.1), outline competing approaches (Section 5.2), and present experimental results (Section 5.3). Results and insights are discussed in the next section.

5.1. Experiment Setting

As a proof-of-concept for our proposed system, we experiment on the publicly available MALib dataset. This comprises 704,128 dependencies between a collection of 56,091 Android GitHub projects to 763 Android third-party libraries. To evaluate recommendation quality, we follow a methodology common in library recommendation research [11,14,16]. In particular, we select all projects with at least 10 dependencies as test ones (these are 31,438 in total), by merit of them comprising enough dependencies to be considered high-quality known ground truth. For these projects, we remove $rm \in \{1, 3, 5\}$ dependencies to emulate the real-world scenario where not all relevant libraries are used and conduct experiments where we use the remaining dependencies to rediscovering the removed ones.

For each approach, the following measures assess the quality of the top $T \in \{5, 10\}$ library recommendations. All measures output values in the range $[0, 1]$, with higher values indicating recommendations closer to ideal ones.

MAP. The mean average precision of the top T recommendations. In detail, given the notation $L_{proj}[i] = \{1$ if the i-th top library recommendation for project *proj* is a true positive, 0 otherwise$\}$, we compute the average precision for each project's top T library recommendations per

$$AP_{proj} = \frac{\sum_{i=1}^{T} L_{proj}[i] \sum_{j=1}^{T} L_{proj}[j] / i}{\sum_{i=1}^{T} L_{proj}[i]}$$

and report its mean across all projects. Average precision provides a more granular understanding than precision by accounting for recommendation order and is thus able to differentiate between recommendation algorithm quality even for large T.

MP. The mean precision of the top T recommendations across all projects. Higher values indicate that there are fewer erroneous library recommendations in the list of top ones. Perfect library recommendations yield MP equal to $\min\{rm/T, 1\}$.

MR. The mean recall of the top T recommendations across all projects. Higher values indicate that there are fewer desired library recommendations (i.e., from those of each project's test set) left out. Perfect library recommendations yield an MR equal to 1.

MF1. The mean F1 score of the top T recommendation across all projects. The F1 score for a project is the harmonic mean between its precision and recall. Then, the mean of all these scores is obtained.

Cov. The coverage of recommendations is the percentage of libraries that reside in the top T recommendation of at least one software project. A coverage value of 1 means that all libraries can be recommended, whereas low percentages indicate approaches that prioritize a few well-known libraries—an undesirable outcome when the goal of recommendation is also to discover fitting non-popular libraries.

5.2. Compared Approaches

In addition to our approach (LibFilter), our experiments assess the following representation learning architectures and graph filters. These are summarized in Table 1 alongside amortized training and inference (making recommendations for one project) times. Time analysis holds for connected dependency graphs and explores terms pertaining to scalability with regards to the numbers of dependencies E and of libraries $I < E$, as well as architectural characteristics, namely the latent representation dimensions *dims*, the number of training *epochs*, and the constant numerical tolerance ϵ of iterative methods computing graph filters. In practice, the number of layers, dimensions and training epochs introduce huge multiplicative terms to running times (in the tens or hundreds order of magnitude each). They could also grow with the number of dependencies, as more effort is required to learn from larger datasets. Thus, even when dimension terms can be removed with parallelized GPU computing, representation training times could scale worse than linearly with the number of mined dependencies.

GRec. A library recommendation approach based on state-of-the-art GNNs for link prediction [16]. It implements convolutional self-attention layers of (7), whose outcomes are concatenated and used as representations. Layers are trained with 10% dropout and comprise 128 latent dimensions and representation matrices inputted to the first layer $H^{(0)}$ are trained end-to-end. We refer to the architecture of the respective paper for more details. This architecture's latent representations need to be retrained to make predictions for new software projects.

LibSeek. A matrix factorization approach [14] that aims to find project and library representations able reconstruct dependency graph adjacency matrices per (1). It was the previous and widely recognized state-of-the-art approaches before GRec and was one of the first to explicitly recognize the diversification of recommendations (i.e., high coverage) as an important goal of library recommendation. Notably, we do not compare against previous works because these have been found to yield a similar or lower recommendation quality across all measures on the dataset we experiment on [14].

LibPPR. Collaborative filtering that employs the personalized PageRank graph filter for recommendation. As described by Bahmani et al. [22], this was once a popular approach. Although it has since been abandoned in favor of GNNs, it is the approach that is closest to ours since it also employs graph filters. Notably, personalized PageRank depends on a diffusion parameter $a \in [0, 1)$, which for smaller values creates lower-pass versions of the graph filter. We follow a random walk with restart formulation that has an equal chance to restart the walks as moving to neighbors and set this parameter to $a = 0.5$. Given that $\frac{1}{1-a} = 2$ is the average length of the random walk processes modeled by personalized PageRank [40], this creates a receptive field that places emphasis on projects and libraries co-used with the query ones, as these lie two hops away from the query ones. We empirically corroborated that this is better-performing than the most widely adopted alternative $a = 0.85$ or even shorter average random walk lengths arising from $a = 0.25$.

LibARW. Collaborative filtering that employs the partially absorbing random walks of (9) for absorption rates $S = \frac{1-a}{a} \mathcal{I}$. This graph filter was proposed [39]. Given that the parameter $a \in (0, 1)$ is equivalent to the one of personalized PageRank, we select $a = 0.5$ for this approach, the same value as LibPPR. We also empirically corroborate that this is performs better than alternatives, such as $a = 0.25$ and $a = 0.85$. Importantly, since LibARW is not our proposed approach, empirical investigation does *not* introduce overtraining bias to experiment results.

LibFilter. Collaborative filtering that employs our proposed symmetric partially absorbing random walks that apply on (9) the absorption rates determined by (10). This approach is a true no-learning one in that it requires no parameter training and no hyperparameter tuning.

Table 1. Overview of compared library recommendation approaches, including training and inference (for one project) times. Recommendation times are bottlenecked by both training and inference.

Approach	Citation	Type	Training Time	Inference Time
LibSeek	[14]	Repr. learning	$O\left(E \cdot dims \cdot layers \cdot epochs\right)$	$O\left(I \cdot dims\right)$
GRec	[16]	Repr. learning	$O\left(E \cdot dims \cdot epochs\right)$	$O\left(I \cdot dims\right)$
LibPPR	[22], this work	Graph filter	—	$O\left(-E \cdot \log \epsilon\right)$
LibARW	[39], this work	Graph filter	—	$O\left(-E \cdot \log \epsilon\right)$
LibFilter	[this work]	Graph filter	—	$O\left(-E \cdot \log \epsilon\right)$

5.3. Results

Table 2 presents the outcome of experimentally evaluating competing approaches. Since GRec and LibSeek follow the same evaluation methodology as we do, we pull evaluation results for these approaches from respective publications. We do not run the publicly available code of GRec and LibSeek to avoid biasing our comparison with lower-quality results arising from post-publication experimental probing by development teams. For instance, we failed to set up GRec's latest published code version to reach the same high evaluation scores as those reported by their paper and found architectural inconsistencies (including different types of layers and activations) between the paper and the code while investigating the issue. Thus, we decided to err on the side of caution and present the better reported values. Graph filters were implemented by building on the filter definition framework provided by the *pygrank* Python package [19] and were run on its *numpy* backend to 10^{-12} mean absolute error numerical tolerance. (*pygrank*'s *numpy* backend implements the graph shift operator by wrapping the C++ code for sparse matrix multiplication and runs faster than the respective operation provided by existing GPU computing frameworks. We ran experiments five times and reported measure averages across runs. Standard deviations are less than 0.007 for coverage and less than 0.002 for other recommendation quality measures and thus facilitate robust pairwise approach comparisons. An implementation of the symmetric absorbing random walk filter and the experiment methodology are publicly available online (https://github.com/maniospas/libFilter accessed on 2 March 2022) .

For all recommendation quality measures aside from MAP, there is a clear evaluation order where GRec is the best approach and is followed by LibFiter (our approach), where the latter lags behind by an at most 5–23% relative decrease that shrinks as more dependencies are omitted from the training graph. Although the two approaches do not always enjoy similar levels of recommendation quality, they can be considered roughly comparable when factoring in the much lower predictive quality of LibSeek and LibPPR. In fact, these last two approaches lag significantly behind, especially in terms of coverage, for which they exhibit near-half or less of GRec. Characteristically, LibFilter lies approximately mid-way between GRec and LibSeek in terms of evaluation measures. Furthermore, it outperforms the other two filter-based alternatives LibPPR and LibARW by a large and small margin, respectively, across all experiments.

With regard to practical deployment, we ran graph filters (LibPPR, LibARW, LibFilter) in a machine with 2.6 GHz CPU base clock and 16GB DDR3 RAM. This extracts library recommendation scores for each project approximately within a fifth of a second—and well within 0.1 second when LibFilter is deployed. By comparison, the out-of-the-box running of the publicly available implementation of GRec on the same machine's GPU with 1680 MHz base clock and 6GB DDR6 graphics memory requires over 5.5 h for training alone (approximately 25 s per training epoch for 800 epochs); this would be the minimum recommendation delay in case of deployment as a query-able service.

Table 2. Comparison of library recommendation approaches.

Approach	Top 5 Recommendations					Top 10 Recommendations					No-Learn
	MP	MR	MF1	MAP	Cov	MP	MR	MF1	MAP	Cov	
Leave out 1 test library per project											
GRec	0.152	0.761	0.254	0.623	0.695	0.083	0.828	0.151	0.636	0.792	x
LibSeek	0.135	0.674	0.225	0.524	0.335	0.076	0.755	0.137	0.535	0.396	x
LibPPR	0.119	0.596	0.199	0.461	0.211	0.072	0.715	0.130	0.477	0.283	✓
LibARW	0.135	0.676	0.226	0.528	0.520	0.077	0.772	0.140	0.541	0.602	✓
LibFilter	0.140	0.700	0.234	0.552	0.544	0.079	0.789	0.143	0.564	0.620	✓
Leave out 3 test libraries per project											
GRec	0.410	0.692	0.514	0.797	0.685	0.234	0.788	0.360	0.761	0.782	x
LibSeek	0.371	0.618	0.464	0.728	0.325	0.216	0.719	0.332	0.697	0.391	x
LibPPR	0.330	0.550	0.413	0.575	0.241	0.207	0.691	0.319	0.509	0.324	✓
LibARW	0.377	0.628	0.471	0.595	0.557	0.224	0.746	0.344	0.535	0.640	✓
LibFilter	0.391	0.652	0.489	0.597	0.579	0.228	0.760	0.351	0.542	0.655	✓
Leave out 5 test libraries per project											
GRec	0.587	0.594	0.590	0.840	0.657	0.361	0.731	0.483	0.786	0.754	x
LibSeek	0.529	0.529	0.529	0.790	0.314	0.329	0.658	0.439	0.740	0.380	x
LibPPR	0.495	0.495	0.495	0.572	0.282	0.329	0.658	0.438	0.470	0.369	✓
LibARW	0.570	0.570	0.570	0.562	0.599	0.357	0.714	0.476	0.474	0.689	✓
LibFilter	0.588	0.588	0.588	0.558	0.602	0.363	0.725	0.484	0.476	0.688	✓

6. Discussion

In this section, we discuss the experiment results and how these can be interpreted within the scope of library recommendation. We also point out promising research directions motivated by our findings, both in automated software engineering and in broader collaborative filtering research. We start by comparing our approach to representation learning techniques (Section 6.1) and assess whether we meet the goal of performing fast library recommendation without lagging excessively far behind in terms of recommendation quality. We also explore the role of filtering memory in improving recommendation algorithms and propose that this direction needs to be explored more thoroughly in the future. Furthermore, based on comparison between graph filter alternatives with different memory mechanisms in their ability to predict relevant libraries to software projects, we propose that searching for new libraries among popular ones is less important than looking at the libraries employed by similar software projects (Section 6.2). Finally, we outline the threats to evaluation validity and describe how these can be addressed when creating real-world systems (Section 6.3).

6.1. Qualitative Approach Comparison

Looking at the experimental results of Table 2 in greater detail, our proposed LibFilter system outperforms the matrix factorization of LibSeek in terms of recommendation quality, which we attribute to the wider receptive field of graph filters that explicitly accounts for co-usage patterns more than one hop away in dependency graphs. On the other hand, LibFilter lags behind the GNN architecture of GRec, which both learns representations and accounts for a wide receptive field. Nonetheless, evaluation in all experiments lies significantly closer to GRec and we consider deviations from the latter small enough for practical deployment sensibilities to play a greater role when choosing which approach to employ. In fact, when dependency graphs have many missing links, as happens for $rm = 5$, our approach catches up in terms of the MP, MR and MF1 measures. Therefore, we argue that the usage of LibFilter should be preferred as an out-of-the-box solution in place of more sophisticated representation-learning alternatives, as the latter need hours instead of fractions of a second to recommend libraries for a new project and would require additional software engineering investigation to determine their viability for the services being developed.

We then point out that the well-established practice of deploying personalized PageRank filters, which we modeled with LibPPR, fails to achieve a similar recommendation quality across all experiments and thus cannot be considered for real-world usage. In fact, it is outperformed by LibSeek to say nothing of the more sophisticated GRec. This result corroborates why collaborative filtering has moved away from graph filters and towards representation learning. However, at least for library recommendation tasks, our experiments suggest that the issue lies less with the inherent power of filters and more with naive structural assumptions (e.g., memory-less random walks) firmly embedded in popular literature, which tend to promote the blind usage of personalized PageRank filters.

Our research escapes from this line of thinking by theorizing that the explicit memory-aware components of partially absorbing random walks can capture dynamics similar to a human-driven library search. Although personalized PageRank is also a type of partially absorbing random walk, it exhibits a memory strongly biased towards node popularity rather than relevance to search outcomes, i.e., an equivalent human search would prioritize remembering and looking at popular libraries (more on this in the next subsection). The importance of search memory for library recommendation is further accentuated if we consider that GRec introduces memory-like constructs in the form of node self-attention terms that multiply incoming representations with those already found in nodes. On the other hand, matrix factorization approaches, such as LibSeek, do not model similar phenomena.

Together, these two findings indicate that, contrary to the popularity-based biasing of results, node-specific memory could be the critical research direction for qualitative collaborative filtering algorithms. Given the success of GNN attention in other link prediction tasks, these findings could also translate to domains beyond library recommendation. Furthermore, our research indicates that the usage of graph filters in link prediction systems should be reconsidered as a viable alternative that can compete at, if not the same, at least comparable levels to GNNs while accommodating practical considerations. In particular, our approach is deployed in the form of a graph filter that can be applied to any structure-based link prediction task, even in other domains where it can potentially remove the need for GNN training. Nonetheless, its efficacy in new tasks should be investigated first.

6.2. Library Popularity and Memorability

Leaving aside representation learning for a moment, the three graph filters we experimented with were derived from partially absorbing random walks and differ only with respect to what type of memory they employ. In particular, LibPPR places higher emphasis on remembering higher-degree nodes, LibARW places the same emphasis on remembering all nodes, and LibFilter performs a type of trade-off between the two. The results indicate that the trade-off yields better recommendations than the other two, thus validating our symmetric principle. Nonetheless, LibARW follows closely behind, which indicates that it is more important for mechanisms remembering relevant libraries to be near-unbiased with respect to popularity, i.e., the number of projects using them.

Looking at this finding from a practical perspective, real-world popularity is only a small indicator of library quality. That is, it is not always worth using popular libraries marginally matching the project at hand. To the contrary, our findings indicate that using highly specialized libraries should be preferred as long as they better fit target tasks—though when suitability is a tie, then selecting the more popular ones to remember is still a valid practice. By extension, we propose that popularity-based metrics (e.g., stars, forks) often used as indicators of potential impact to development communities could be misleading by themselves and new qualitative-based metrics should be introduced.

6.3. Threats to Validity

Before concluding this work, we outline potential threats to our research's validity.

First, we used popular measures to assess the quality of library recommendations. Previous related works have often performed developer surveys to corroborate the efficacy of experiments, for example, by emailing developers with libraries recommended for their projects with multiple systems and obtaining feedback on whether these could be of actual interest. In this work, we did not do this. However, we point out that developer feedback for the assessment of previous systems (including those we compare our approach against) has shown strong correlation between practical usefulness and the evaluation measures we employ [14,16]. Therefore, recreating the same small-scale studies could be considered redundant, especially since evaluation measure values lay in interpolate-able points between existing approaches. That is, our approach does not further improve recommendation quality but changes computational costs to be scalable. Thus, there exist no reasonable concerns over employed metrics failing to capture a practical impact.

When interpreting training and inference time measures, we caution that different approaches integrate different computing frameworks and exact numbers could be subject to change depending on the hardware or algorithmic optimizations available. For example, we run *pygrank* on its *numpy* backend because at the time of writing, it is faster than GPU computing for sparse matrix multiplication, but this could change in the future. Nonetheless, we expect that the amortized running times presented in Table 1 will yield the similar scalability of approaches. In this case, the driving criteria of algorithmic comparisons are still the training vs. no-training paradigm.

In a related vein, this work considers representation learning to be so time-consuming that architectures cannot be quickly and repeatedly retrained. This is not likely to change in the foreseeable future, especially since data tend to grow at faster rates than computational resources. However, one promising alternative would be to perform the warm-start training of GNNs to answer queries, for example, with streaming training principles [41]. Whether this would be useful for a library recommendation is yet unknown, for instance, due to the often degraded predictive quality of stream learning, or due to the local optimal regions drifting substantially so that minor representation tweaks are not sufficient.

Another threat to validity comes from experimenting on only one dataset. Although evaluation on this dataset is the gold standard in collaborative library recommendation literature, we stress that competing approaches could exhibit different efficacies if applied to different types of dependency graphs, such as the library-to-library dependency graphs, which are not bipartite. Thus, we point out that future research could also move towards benchmarking approaches on multiple datasets. We stress that this concern is shared across the whole collaborative library recommendation literature and not only our approach. For the time being, we propose that developers of real-world library recommendation services perform experimental probings to verify that selected recommendation algorithms replicate promised quality benefits on their own data, for example, with the evaluation methodology described in this work.

Finally, in line with previous research, we follow a collaborative recommendation approach. This makes use of known project–library dependencies to recommend more links but ignores real-world semantics such as project names or descriptions. Enriching library recommendations with semantics is a promising direction for future work as it could potentially procure recommendations with no known dependencies, for example to bootstrap development in unfamiliar domains. However, additional exploration is needed to (a) formulate how to extend graph inference on content features without resorting to the end-to-end training of latent representations; and (b) verify whether semantics are useful for library recommendation.

7. Conclusions

In this work, we discussed the problem of recommending library dependencies for new software projects based on co-usage patterns in other projects. For this task, we recognized that existing representation learning approaches exhibit the practical limitation of needing to retrain to make recommendations for new projects, hindering widespread adoption, and explained that no-learning project–library dependency graph inference circumvents this shortcoming. We proposed that graph filters match this paradigm and introduced a novel variation of partially absorbing random walk filters, which we theorized to emulate human-driven library discovery by modeling the memorization of libraries and projects similar to query ones. To show our approach's efficacy, we experimented in a real-world dependency graph of Android project third-party library dependencies, where we found that it did not lag significantly behind state-of-the-art representation learning, where the latter introduces long recommendation delays when deployed to factual systems.

Author Contributions: Funding acquisition, A.S.; investigation, E.K.; methodology, E.K. and A.S.; project administration, A.S.; software, E.K.; supervision, A.S.; visualization, E.K. and A.S.; writing— original draft, E.K.; writing—review and editing, A.S. All authors have read and agreed to the published version of the manuscript.

Funding: This research has been co-financed by the European Regional Development Fund of the European Union and Greek national funds through the Operational Program Competitiveness, Entrepreneurship and Innovation, under the call RESEARCH-CREATE-INNOVATE (project code: T2EΔK-00550).

Data Availability Statement: The MALib dataset analyzed in this study was imported from its publicly available repository here: https://github.com/malibdata/MALib-Dataset, accessed on 2 March 2022.

Conflicts of Interest: The funders had no role in the design of the study, in the collection, analysis, or interpretation of data, in the writing of the manuscript, or in the decision to publish the results.

Abbreviations

The following abbreviations are used in this manuscript:

Cov	Coverage
MAP	Mean Average Precision
MF1	Mean F1 Score
MP	Mean Precision
MR	Mean Recall
GNN	Graph Neural Network
GPU	Graphics Processing Unit

References

1. Nerur, S.; Balijepally, V. Theoretical reflections on agile development methodologies. *Commun. ACM* **2007**, *50*, 79–83. [CrossRef]
2. Miller, F.P.; Vandome, A.F.; McBrewster, J. *Apache Maven*; Alpha Press: Indianapolis, IN, USA, 2010.
3. Python Package Index—PyPI. Python Software Foundation. Available online: https://pypi.org (accessed on 2 March 2022).
4. npm. npm, Inc. Available online: https://www.npmjs.com (accessed on 2 March 2022).
5. Raemaekers, S.; Van Deursen, A.; Visser, J. The maven repository dataset of metrics, changes, and dependencies. In Proceedings of the 2013 10th Working Conference on Mining Software Repositories (MSR), San Francisco, CA, USA, 18–19 May 2013 ; IEEE Computer Society: Washington, DC, USA, 2013; pp. 221–224.
6. Li, Z.; Avgeriou, P.; Liang, P. A systematic mapping study on technical debt and its management. *J. Syst. Softw.* **2015**, *101*, 193–220. [CrossRef]
7. He, X.; Liao, L.; Zhang, H.; Nie, L.; Hu, X.; Chua, T.S. Neural collaborative filtering. In Proceedings of the 26th International Conference on World Wide Web, Perth, Australia, 3–7 May 2017; pp. 173–182.
8. Barbosa, E.A.; Garcia, A. Global-aware recommendations for repairing violations in exception handling. *IEEE Trans. Softw. Eng.* **2017**, *44*, 855–873. [CrossRef]
9. Huang, Q.; Xia, X.; Xing, Z.; Lo, D.; Wang, X. API method recommendation without worrying about the task-API knowledge gap. In Proceedings of the 2018 33rd IEEE/ACM International Conference on Automated Software Engineering (ASE), Montpellier, France, 3–7 September 2018; IEEE: Piscataway, NJ, USA, 2018; pp. 293–304.
10. Ichii, M.; Hayase, Y.; Yokomori, R.; Yamamoto, T.; Inoue, K. Software component recommendation using collaborative filtering. In Proceedings of the 2009 ICSE Workshop on Search-Driven Development-Users, Infrastructure, Tools and Evaluation, Vancouver, BC, Canada, 16 May 2009; IEEE: Piscataway, NJ, USA, 2009; pp. 17–20.
11. Thung, F.; Lo, D.; Lawall, J. Automated library recommendation. In Proceedings of the 2013 20th Working conference on reverse engineering (WCRE), Koblenz, Germany, 14–17 October 2013; IEEE: Piscataway, NJ, USA, 2013; pp. 182–191.

12. Ouni, A.; Kula, R.G.; Kessentini, M.; Ishio, T.; German, D.M.; Inoue, K. Search-based software library recommendation using multi-objective optimization. *Inf. Softw. Technol.* **2017**, *83*, 55–75. [CrossRef]
13. Su, X.; Khoshgoftaar, T.M. A survey of collaborative filtering techniques. *Adv. Artif. Intell.* **2009**, *2009*, 421425. [CrossRef]
14. He, Q.; Li, B.; Chen, F.; Grundy, J.; Xia, X.; Yang, Y. Diversified third-party library prediction for mobile app development. *IEEE Trans. Softw. Eng.* **2020**, *48*, 150–165. [CrossRef]
15. Bottou, L. Large-scale machine learning with stochastic gradient descent. In Proceedings of the COMPSTAT'2010: 19th International Conference on Computational Statistics, Paris, France, 22–27 August 2010; Keynote, Invited and Contributed Papers; Springer: Berlin/Heidelberg, Germany, 2010; pp. 177–186.
16. Li, B.; He, Q.; Chen, F.; Xia, X.; Li, L.; Grundy, J.; Yang, Y. Embedding app-library graph for neural third party library recommendation. In Proceedings of the 29th ACM Joint Meeting on European Software Engineering Conference and Symposium on the Foundations of Software Engineering, Athens, Greece, 23–28 August 2021; pp. 466–477.
17. Yan, D.; Tang, T.; Xie, W.; Zhang, Y.; He, Q. Session-based Social and Dependency-aware Software Recommendation. *arXiv* **2021**, arXiv:2103.06109.
18. Ortega, A.; Frossard, P.; Kovačević, J.; Moura, J.M.; Vandergheynst, P. Graph signal processing: Overview, challenges, and applications. *Proc. IEEE* **2018**, *106*, 808–828. [CrossRef]
19. Krasanakis, E.; Papadopoulos, S.; Kompatsiaris, I.; Symeonidis, A. pygrank: A Python Package for Graph Node Ranking. *arXiv* **2021**, arXiv:2110.09274.
20. Page, L.; Brin, S.; Motwani, R.; Winograd, T. *The PageRank Citation Ranking: Bringing Order to the Web*; Technical Report; Stanford InfoLab: Stanford, CA, USA, 1999.
21. Andersen, R.; Chung, F.; Lang, K. Local graph partitioning using pagerank vectors. In Proceedings of the 2006 47th Annual IEEE Symposium on Foundations of Computer Science (FOCS'06), Berkeley, CA, USA, 21–24 October 2006; IEEE: Piscataway, NJ, USA, 2006; pp. 475–486.
22. Bahmani, B.; Chowdhury, A.; Goel, A. Fast incremental and personalized pagerank. *arXiv* **2010**, arXiv:1006.2880.
23. Chung, F. The heat kernel as the pagerank of a graph. *Proc. Natl. Acad. Sci. USA* **2007**, *104*, 19735–19740. [CrossRef]
24. Kloster, K.; Gleich, D.F. Heat kernel based community detection. In Proceedings of the 20th ACM SIGKDD International Conference on Knowledge Discovery and Data Mining, New York, NY, USA, 24–27 August 2014; pp. 1386–1395.
25. Wu, Z.; Pan, S.; Chen, F.; Long, G.; Zhang, C.; Philip, S.Y. A comprehensive survey on graph neural networks. *IEEE Trans. Neural Netw. Learn. Syst.* **2020**, *32*, 4–24. [CrossRef] [PubMed]
26. Zhang, Z.; Cui, P.; Zhu, W. Deep learning on graphs: A survey. *IEEE Trans. Knowl. Data Eng.* **2020**, *34*, 249–270. [CrossRef]
27. Wang, X.; He, X.; Wang, M.; Feng, F.; Chua, T.S. Neural graph collaborative filtering. In Proceedings of the 42nd international ACM SIGIR conference on Research and development in Information Retrieval, Paris, French, 21–25 July 2019; pp. 165–174.
28. Kipf, T.N.; Welling, M. Semi-supervised classification with graph convolutional networks. *arXiv* **2016**, arXiv:1609.02907.
29. Agarap, A.F. Deep learning using rectified linear units (relu). *arXiv* **2018**, arXiv:1803.08375.
30. Hamilton, W.L.; Ying, R.; Leskovec, J. Inductive representation learning on large graphs. In Proceedings of the 31st International Conference on Neural Information Processing Systems, Long Beach, CA, USA, 4–9 December 2017; pp. 1025–1035.
31. Klicpera, J.; Bojchevski, A.; Günnemann, S. Predict then propagate: Graph neural networks meet personalized pagerank. *arXiv* **2018**, arXiv:1810.05997.
32. Chen, M.; Wei, Z.; Huang, Z.; Ding, B.; Li, Y. Simple and deep graph convolutional networks. In Proceedings of the International Conference on Machine Learning. PMLR, Virtual Event, 13–18 July 2020; pp. 1725–1735.
33. Adamczyk, P.; Smith, P.H.; Johnson, R.E.; Hafiz, M. Rest and web services: In theory and in practice. In *REST: From Research to Practice*; Springer: Berlin/Heidelberg, Germany, 2011; pp. 35–57.
34. Gunasekaran, A. Agile manufacturing: A framework for research and development. *Int. J. Prod. Econ.* **1999**, *62*, 87–105. [CrossRef]
35. Dong, H.; Chen, J.; Feng, F.; He, X.; Bi, S.; Ding, Z.; Cui, P. On the equivalence of decoupled graph convolution network and label propagation. In Proceedings of the Web Conference 2021, New York, NY, USA, 19–23 April 2021; pp. 3651–3662.
36. Yang, F.; Zhang, H.; Tao, S.; Hao, S. Graph representation learning via simple jumping knowledge networks. *Appl. Intell.* **2022**, 1–19. [CrossRef]
37. Gleich, D.F.; Seshadhri, C. Vertex neighborhoods, low conductance cuts, and good seeds for local community methods. In Proceedings of the 18th ACM SIGKDD International Conference on Knowledge Discovery and Data Mining, Beijing, China, 12–16 August 2012; pp. 597–605.
38. Tong, H.; Faloutsos, C.; Pan, J.Y. Fast random walk with restart and its applications. In Proceedings of the Sixth International Conference on Data Mining (ICDM'06), Hong Kong, China, 18–22 December 2006; IEEE: Piscataway, NJ, USA, 2006; pp. 613–622.
39. Wu, X.M.; Li, Z.; So, A.; Wright, J.; Chang, S.F. Learning with partially absorbing random walks. *Adv. Neural Inf. Process. Syst.* **2012**, *25*, 3077–3085 .
40. Krasanakis, E.; Papadopoulos, S.; Kompatsiaris, I. Stopping personalized PageRank without an error tolerance parameter. In Proceedings of the 2020 IEEE/ACM International Conference on Advances in Social Networks Analysis and Mining (ASONAM), Hague, The Netherlands, 7–10 December 2020; IEEE: Piscataway, NJ, USA, 2020; pp. 242–249.
41. Wang, J.; Song, G.; Wu, Y.; Wang, L. Streaming graph neural networks via continual learning. In Proceedings of the 29th ACM International Conference on Information & Knowledge Management, Virtual, 19–23 October 2020; pp. 1515–1524.

future internet

Article

Ontology-Based Feature Selection: A Survey

Konstantinos Sikelis, George E. Tsekouras * and Konstantinos Kotis

Department of Cultural Technology and Communications, University of the Aegean, 811 00 Mitilini, Greece; cti20004@ct.aegean.gr (K.S.); kotis@aegean.gr (K.K.)
* Correspondence: gtsek@ct.aegean.gr; Tel.: +30-22-510-36631

Abstract: The Semantic Web emerged as an extension to the traditional Web, adding meaning (semantics) to a distributed Web of structured and linked information. At its core, the concept of ontology provides the means to semantically describe and structure information, and expose it to software and human agents in a machine and human-readable form. For software agents to be realized, it is crucial to develop powerful artificial intelligence and machine-learning techniques, able to extract knowledge from information sources, and represent it in the underlying ontology. This survey aims to provide insight into key aspects of ontology-based knowledge extraction from various sources such as text, databases, and human expertise, realized in the realm of feature selection. First, common classification and feature selection algorithms are presented. Then, selected approaches, which utilize ontologies to represent features and perform feature selection and classification, are described. The selective and representative approaches span diverse application domains, such as document classification, opinion mining, manufacturing, recommendation systems, urban management, information security systems, and demonstrate the feasibility and applicability of such methods. This survey, in addition to the criteria-based presentation of related works, contributes a number of open issues and challenges related to this still active research topic.

Keywords: feature selection; ontology; text classification; machine-learning

Citation: Sikelis, K.; Tsekouras, G.E.; Kotis, K. Ontology-Based Feature Selection: A Survey. *Future Internet* **2021**, *13*, 158. https://doi.org/10.3390/fi13060158

Academic Editor: Davide Tosi

Received: 5 May 2021
Accepted: 13 June 2021
Published: 18 June 2021

Publisher's Note: MDPI stays neutral with regard to jurisdictional claims in published maps and institutional affiliations.

1. Introduction

The vast amount of information available in the continuously expanding Web by far exceeds human processing capabilities. This problem has been transformed to the research question of whether it is possible to develop methods and tools that will automate the retrieval of information and the extraction of knowledge from Web repositories. The Semantic Web emerged as a technological solution to this problem. In its essence, it is an extension to the traditional Web, where content is now represented in such a way that machines are able to process it (machine-processable) and infer new knowledge out of it. The goal is to alleviate the limitations of current knowledge engineering technology with respect to searching, extracting, maintaining, uncovering, and viewing information, supporting advanced knowledge-based systems. Within the Semantic Web framework, information is organized in conceptual spaces according to its meaning. Automated tools search for inconsistencies and ensure content integrity. Keyword-based search is replaced by knowledge extraction through semantic query answering.

The recent development of the Semantic Web enables the systematic representation of vast amounts of knowledge within an ontological framework. An ontology is a formal and explicit description of shared and agreed knowledge shaped as a set of concepts (and their properties) within a domain of discourse, and binary relationships that hold among them. The ontological model provides a rich set of axioms to link pieces of information, and enables automated reasoning to infer knowledge that has not been explicitly asserted before.

In many cases, reasoning with knowledge can be cast as a data classification task. An important step towards an accurate and efficient classification is feature selection.

Consequently, identification of high-quality features from an ontological hierarchy plays a significant role in the ability to extract information from documents.

The main research domain where ontologies have been employed in terms of selecting specific features is text classification, where predefined categories are associated with free-text unstructured documents based on their content. The continuous increase of volumes of text documents on the Web makes text classification an important tool for searching information. Due to their enormous scale in terms of the number of classes, training examples, features, and feature dependencies, text classification applications present considerable research challenges.

In standard feature selection approaches, feature representation and selection are the main tasks prior to the classification, whereas in the ontology-based feature selection approaches, the task of feature extraction and selection from the input data based on a data-to-ontology mapping is required.

This paper presents related work on the problem of feature representation and selection based on ontologies in the context of knowledge extraction from documents, databases, and human expertise. Beyond important issues related to the volume, velocity, variety, and veracity (4 V) of the Web of (Big) data, the presented work has been motivated by a number of open issues and challenges that keep this research topic still active, especially in the era of Knowledge Graphs (KG) and Linked Open Data (LOD), where bias at different levels (data, schema, reasoning) may cause the development of "unfair" models in different application domains. Furthermore, developing ontology-based feature selection methods for achieving real-time analysis and prediction regarding high-dimensional datasets remains a key challenge. Several research issues related to the use of ontologies in feature selection for classification problems are investigated. The first issue refers to the application areas of ontology-based feature selection. This survey concentrates on a wide range of application areas such as document classification, opinion mining, selection of manufacturing processes, recommendation systems, urban management, and information security, where certain algorithmic structures are discussed, depending on the application framework. The second issue investigates the motivations for building an ontology in order to perform feature selection. Regarding this issue, the current analysis suggests that the above motivations are mainly based on the fact that an ontology provides structured knowledge representation as well as measures of semantic similarity. The former renders the ontology reusable, while the later determines the applicability of the ontology in multiple domains and algorithmic frameworks. Finally, other issues are related to the nature of the algorithmic frameworks and the types of the ontologies used. This survey indicates a wide diversity on the feature selection schemes, where the most common mechanisms are based on filter-based methods, and different domain ontologies such as existing ones or custom, which can be either crisp or fuzzy.

The structure of this survey paper is as follows. In Sections 2 and 3, preliminaries on data classification and feature selection methods are presented. In Section 4, the concept of ontology as a building block of the Semantic Web is introduced. In Section 5, ontology-based feature selection is presented, along with related works organized in application domains and other criteria. In Section 6 open issues and challenges are discussed. Finally, Section 7 concludes this survey.

2. Classification Methods

One of the most common applications of machine learning is data classification. In essence, data classification investigates the relations between feature variables (i.e., inputs) and output variables. Classification methods have been used in a broad range of applications such as customer target marketing [1,2], medical disease diagnosis [3–5], speech and handwriting recognition [6–9], multimedia data analysis [10,11], biological data analysis [12], document categorization and filtering [13,14], and social network analysis [15–17]. Classification algorithms typically contain two steps, the learning step and the testing step. The first one constructs the classification model, while the second evaluates it by assigning class labels to unlabeled data. A close relative to the classification problem is data clustering [18,19]. Clustering is the task of dividing a population of data points into a number of groups, such that the members of the same group are in some sense similar to each other and dissimilar to the data points in other groups. In general the classification task is based on supervised learning, whereas clustering is based on unsupervised learning.

A plethora of methods can be used for data classification. Some of the most common are probabilistic methods [20–22], decision trees [23–25], rule-based methods [26–28], support vector machine methods [29,30], instance-based methods, and neural networks [31,32].

2.1. Probabilistic Data Classification

Probabilistic methods are based on two probabilities, namely a prior probability, which is derived from the training data, and a posterior probability that a test instance belongs to a particular class. There are two approaches for the estimation of the posterior probability. In the first approach, called generative, the training dataset is used to determine the class probabilities and class-conditional probabilities and the Bayes theorem is employed to calculate the posterior probability. In the second approach, called discriminative, the training dataset is used to identify a direct mapping of a test instance onto a class.

A widely used example of generative model is the naive Bayes classifier [31,32], while a popular discriminative classifier is the logistic regression [31].

2.2. Decision Tree Data Classification

In decision tree classification [23–25], data are recursively split into smaller subsets until all formed subsets exhibit class purity, i.e., all members of each subset are sufficiently homogeneous and belong to the same unique class. In order to optimize the decision tree, an impurity measure is employed and the optimal splitting rule at each node is determined by maximizing the impurity decrease due to the split. A commonly used function for this purpose is the Shannon entropy.

An extension to decision tree classification is the Random Forest (RF) algorithm [33]. This algorithm trains a large set of decision trees and combines their predictive ability in a single classifier. The RF classifier belongs to a broader family of methods called ensemble learning [31].

2.3. Rule-Based Data Classification

A classification method closely related to decision trees is called rule-based classification [26–28]. Essentially, all paths in a decision tree represent rules, which map test instances to different classes. However, for rule-based methods the classification rules are not required to be disjointed, rather they are allowed to overlap. Rules can be extracted either directly from data (rule induction) or built indirectly from other classification models.

2.4. Associative Classification

A novel family of algorithms that aim at mining classification rules indirectly, is the so called associative classification [34]. Associations are interesting relations between variables in large datasets. Association rules can quantify such relations by means of constraints on measures of significance or interest. The constraints come in the form of minimum threshold values of support and confidence. In the training phase, an associative classifier, mines a set of Class Association Rules (CARs) from the training data. The mined CARs are used to build the classification model according to some strategy such as applying the strongest rule, selecting a subset of rules, forming a combination of rules, or using rules as features.

2.5. Support Vector Machines

Support vector machine [35] classifiers are generally defined for binary classification tasks. Intuitively, they attempt to draw a decision boundary between the data items of two classes, according to some optimality criterion. A common criterion employed by SVM is that the decision surface must be far away from the data points. The separation degree can be estimated in terms of the distance from the decision surface to the closest data points. Such data points are called support vectors.

Finding the maximum margin hyperplane is a quadratic optimization problem [36]. In case the training data are not linearly separable, slack variables can be introduced in the formulation to allow some training instances to violate the support vector constraint, i.e., they are allowed to be on the "other" side of the support vector from the one that corresponds to their class.

2.6. Artificial Neural Networks Data Classification

Artificial neural networks have been proven to be powerful classifiers [32]. They attempt to mimic the human brain by means of an interconnected network of simple computational units, called neurons. Neurons are functions that map an input feature vector to an output value according to predefined weights. These weights express the influence of each feature over the output of the neuron and are learned during the training phase. A typical tool to perform the training process is the back-propagation algorithm. Back-propagation uses the chain rule to compute the derivative of the error (loss function) with respect to the network's parameters, while gradient-descent-based methods (e.g., stochastic gradient descent) are implemented to find the appropriate weight values.

2.7. Instance-Based Data Classification

Instance-based classifiers do not build any approximation models, rather they simply store the training records [37]. When a query is submitted, the system uses a distance function to extract, from the training data set, those records that are most similar to the test instance. Label assignment is performed based on the extracted subset. Common instance-based classifiers are the K-Nearest Neighbor (KNN), kernel machines, radial basis functions neural networks, etc. [38]. A generalization of instance-based learning is lazy learning, where training examples in the neighborhood of the test instance are used to train a locally optimal classifier. The field of classification is vast and still in its infancy. For an excellent in depth discussion on classification methods, the curious reader is referred to [31].

3. Feature Selection

The first step towards successful classification is to define the features that will be input to the classifier. This process is called Feature Engineering (FE) and encompasses algorithms for generating features from raw data (feature generation), transforming existing features (feature transformation), selecting most important features (feature selection), understanding feature behavior (feature analysis), and determining feature importance (feature evaluation) [39].

Feature selection is well studied under the framework of FE. An increasing number of dimensions in the feature space results in exponential expansion of the computational cost. This issue is directly related to the problem of the curse of dimensionality. Furthermore as the volume of feature space increases, it becomes sparsely populated and even close data points may be driven apart from irrelevant data, thus appearing as far away as unrelated data points. This will increase overfitting and reduce the accuracy of the classifier. Restricting the used features to only those that are strictly relevant to the target classes results in improved interpretability of the model

The feature selection process attempts to remedy these issues by identifying features that can be excluded without adversely affecting the classification outcome. Feature selection is closely related to feature extraction. The main difference is that while feature selection maintains the physical meaning of the retained features, feature extraction attempts to reduce the number of dimensions by mapping the physical feature space on a new mathematical space.

Feature selection can be supervised, unsupervised, or semi-supervised. Supervised methods consider the classification information and use measures to quantify the contribution of each feature to the total information, thus keeping only the most important ones. Unsupervised methods attempt to remove redundant features in two steps. First, features are clustered into groups, using some measure of similarity, and then the features with the strongest correlations to the other features in the same group are retained as the representatives of the group. Identification and removal of irrelevant features is more difficult and abstract and depends on some heuristic of relevance or interestingness. To devise such heuristics, researchers have employed several performance indices namely, category utility, entropy, scatter separability, and maximum likelihood [40]. Semi-supervised feature selection addresses the case when both a large set of unlabeled and a small set of labeled data are available. The idea is to use the supervised class-based clustering of features in the small dataset as constraint for the unsupervised locality-based clustering of the features in the large dataset.

Depending on whether and how they use the classification system, feature selection algorithms are divided into three categories, namely filters, wrappers, and embedded models.

3.1. Filter Models

Filter models determine subsets of features to perform pre-processing, independently of the chosen classifier. In the first step, features are analyzed and ranked on the basis of how they correlate to the target classes. This analysis can either consider features separately and perform ranking independently of the feature space (univariate), or evaluate groups of features (multivariate). Multivariate analysis has the advantage that interactions between features are considered during the selection process. In the second step, the highest ranked (i.e., scored) features constitute the final input variables of the classifier.

Some of the most common evaluation metrics that have been used for ranking and filtering are Chi-square, ANOVA, Fisher score, Pearson correlation coefficient, and mutual information [39–41].

Chi-Square: The χ^2 correlation uses the contingency table of a feature target-pair to evaluate the likelihood that a selected feature and a target class are correlated. The contingency table shows the distribution of one variable (the feature) in rows and another (the target) in columns. Based on the entries, the observed values are calculated under the assumption that the variables are independent (null hypothesis); the expected values are then derived. Small values of χ^2 show that the expected values are close to the observed values, thus the null hypothesis stands. On the contrary, high values show strong correlation between the feature and the target value.

ANOVA: A metric related to χ^2 is analysis of variance. It tests whether several groups are similar or different by comparing their means and variances, and returns an F-statistic, which can be used for feature selection. The idea is that a feature where each of its possible values corresponds to a different target class, will be a useful predictor.

Fisher Score: It is based on the intuition that effective feature combinations should result in similar values regarding instances in the same class, and much different values regarding instances from different classes.

Pearson Correlation Coefficient: It is used as a measure for quantifying linear dependence between a feature variable X_i and a target variable Y_k. It ranges from -1 (perfect anti-correlation) to 1 (perfect correlation).

Mutual Information: The information gain metric provides a method of measuring the dependence between the ith feature and the target classes $\vec{c} = [c_1, c_2, \ldots, c_k]$, as the decrease in total entropy, namely $IG(f_i, \vec{c}) = H(f_i) - H(f_i|\vec{c})$, where $H(f_i)$ is the entropy of f_i and $H(f_i|\vec{c})$ the entropy of f_i after observing $\vec{c}$. High information gain indicates that the selected feature is relevant. IG has been extended to account for feature correlation and redundancy. Other MI metrics are Gini impurity and minimum-redundancy–maximum-relevance.

3.2. Wrapper Models

Filter models select features based on their statistical similarities to a target variable. Wrapper methods take a different approach and use a pre-selected classifier as a way to evaluate the accuracy of the classification task for a specific feature subset. A wrapper algorithm consists of three components, namely a feature search component, a feature evaluation component, and a classifier [39,40]. At each step, the search component generates a subset of features that will be evaluated for the classification task. When the total number of features is small, it is possible to test all possible feature combinations. However, this approach, known as SUBSET, becomes quickly computationally intractable.

Greedy search methods overcome this problem by using a heuristic rule to guide the subset generation [42,43]. In particular, forward selection starts with an empty set and evaluates the classification accuracy of each feature separately. The best feature initializes the set. In the subsequent iterations, the current set is combined with each of the remaining features and the union is tested for its classification accuracy. The feature producing the best classification is added permanently to the selected features and the process is repeated until the number of features reaches a threshold or none of the remaining features improve the classification. On the other hand, backward elimination starts with all features. At each iteration, all features in the set are removed one by one and the resulting classification is evaluated. The feature affecting the classification the least, is removed from the list. Finally, bidirectional search starts with an empty set (expanding set) and a set with all features (shrinking set). At each iteration, first a feature is forward selected and added to the expanding set with the constraint that the added feature exists in the shrinking set. Then a feature is backward eliminated from the shrinking set with the constraint that it has not already been added in the expanding set.

Many more strategies have been used to search the feature space, such as branch-and-bound, simulated annealing, and genetic algorithms [42,43]. Branch-and-bound uses depth-search to traverse the feature subset tree, pruning those branches that have worse classification score than the score of an already traversed fully expanded branch. Simulated annealing and genetic algorithms encode the selected features in a binary vector. At each step, offspring vectors, representing different combinations of features, are generated and tested for their accuracy. A common technique for performance assessment is k-fold cross-validation. The training data are split into k sets and the classification task is performed k times, using at each iteration one set as the validation set and the remaining k-1 sets for training.

3.3. Embedded Methods

Filter methods are cheap, but selected features do not consider the biases of the classifiers. Wrapper methods select features tailored to a given classifier, but have to run the training phase many times, hence they are very expensive [42,43]. Embedded methods combine the advantages of both filters and wrappers by integrating feature selection in the training process. For example, pruning in decision trees and rule-based classifiers is a built-in mechanism to select features. In another family of classification methods, the change in the loss function incurred by changes in the selected features, can be either exactly computed or approximated, without the need to retrain the model for each candidate variable. Combined with greedy search strategies, this approach allows for efficient feature selection (e.g., RFE/SVM, Gram–Schmidt/LLS). A third type of embedded methods are regularization methods and apply to classifiers where weight coefficients are assigned to features (e.g., SVM or logistic regression). In this case, the feature selection task is cast as an optimization problem with two components, namely maximization of goodness-of-fit and minimization of the number of variables. The latter condition is achieved by forcing weights to be small or exactly zero. Features with coefficients close to zero are removed. Specifically, the feature weight vector is defined as in [42,43].

Many more feature selection algorithms and variations can be found in the literature. Due to its significance in the classification task, feature selection, and feature engineering in general, is a highly active field of research. For an in-depth presentation, the interested reader is referred to [39–41]. Comprehensive reviews can be found in [42,43].

4. Ontologies

The enormous amount of information available in the continuously expanding Web by far exceeds human processing capabilities. This gave rise to the question of whether it is possible to build tools that will automate information retrieval and knowledge extraction from the Web repository. The Semantic Web emerged as a proposed solution to this problem. In its essence, it is an extension to the Web, in which content is represented in such a way that machines are able to process it and infer new knowledge from it. Its purpose is to alleviate the limitations of current knowledge engineering technology with respect to searching, extracting, maintaining, uncovering and viewing information, and support advanced knowledge-based systems. Within the Semantic Web framework, information is organized in conceptual spaces according to its meaning. Automated tools search for inconsistencies and ensure content integrity. Keyword-based search is replaced by knowledge extraction through query answering.

In order to realize its vision, the Semantic Web does not rely on "exotic" intelligent technology, where agents are able to mimic humans in understanding the predominant HTML content. Rather it approaches the problem from the Web page side. Specifically, it requires Web pages to contain informative (semantic) annotations about their content. These semantics (metadata) enable software to process information without the need to "understand" it. The eXtensible Markup Language (XML) was a first step towards this goal. Nowadays, the Resource Description Framework (RDF), RDF Scheme (RDFS) and the Web Ontology Language (OWL) are the main technologies that drive the implementation of the Semantic Web.

In general, ontologies are the basic building blocks for inference techniques on the Semantic Web. As stated in W3C's OWL Requirements Documents [44]: "An ontology defines the terms used to describe and represent an area of knowledge". Ontological terms are concepts and properties which capture the knowledge of a domain area. Concepts are organized in a hierarchy that expresses the relationships among them by means of superclasses representing higher level concepts, and subclasses representing specific (constrained) concepts. Properties are of two types: those that describe attributes (features) of the concepts, and those that introduce binary relations between the concepts. An example ontology is depicted in Figure 1.

In order to succeed in the goal to express knowledge in a machine-processable way, an ontology has to exhibit certain characteristics, namely abstractness, preciseness, explicitness, consensus, and domain specificity. An ontology is abstract when it specifies knowledge in a conceptual way. Instead of making statements about specific occurrences of individuals, it tries to cover situations in a conceptual way. Ontologies are expressed in a knowledge representation language that is grounded on formal semantics, i.e., it describes the knowledge rigorously and precisely. Such semantics do not refer to subjective intuitions, nor are they open to different interpretations. Furthermore, knowledge is stated explicitly. Notions that are not directly included in the ontology are not part of the conceptualization it captures. In addition, an ontology reflects a common understanding of domain concepts within a community. In this sense, a prerequisite of an ontology is the existence of social consensus. Finally, it targets a specific domain of interest. The more refined the scope of the domain, the more effective an ontology can be at capturing the details rather than covering a broad range of related topics.

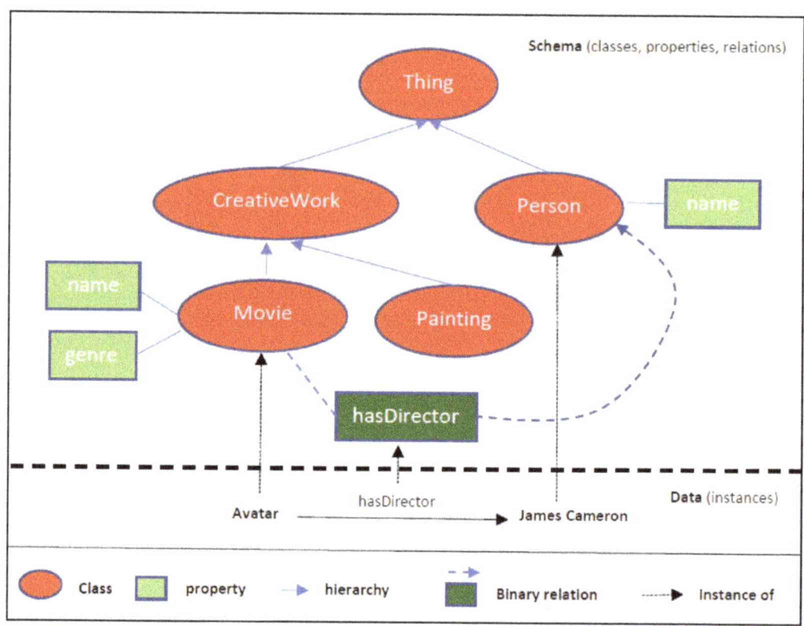

Figure 1. Example ontology.

The most popular language for engineering ontologies is OWL [45]. OWL (and the latest iteration: OWL2) defines constructs, namely classes, associated properties, and binary relationships between those classes, which can be used to create domain vocabularies along with constructs for expressiveness (e.g., cardinalities, unions, intersections), thus enabling the modeling of complex and rich axioms. There are many tools available that support the engineering of OWL ontologies (e.g., Protégé, TopBraid Composer) and OWL-based reasoning (e.g., Pellet, HermiT). Ontology engineering is an active topic and a growing number of fully developed domain and generic/upper ontologies are already publicly available, such as the Dublin Core (DC) [46], the Friend Of A Friend (FOAF) [47], Gene Ontology (GO) [48], Schema.org [49], to name a few. An extensive list of ontologies and related ontology engineering methodologies have been recently published in Kotis et al. [50].

The Semantic Web is vast and combines many areas of research and technological advances. A comprehensive introduction can be found in [51,52]. The interested reader can find a detailed presentation of Semantic Web technologies in [53], and analytical review of semantic annotation of web services in [54].

5. Ontology-Based Feature Selection

In standard feature selection approaches the pipeline of tasks (Figure 2a) include features representation prior to selection, whereas in the ontology-based feature selection pipeline there is need to first extract the related features from the input data (after prepossessing) according to a utilized ontology (mapping) and then select those features that are more suitable for the classification task (Figure 2b).

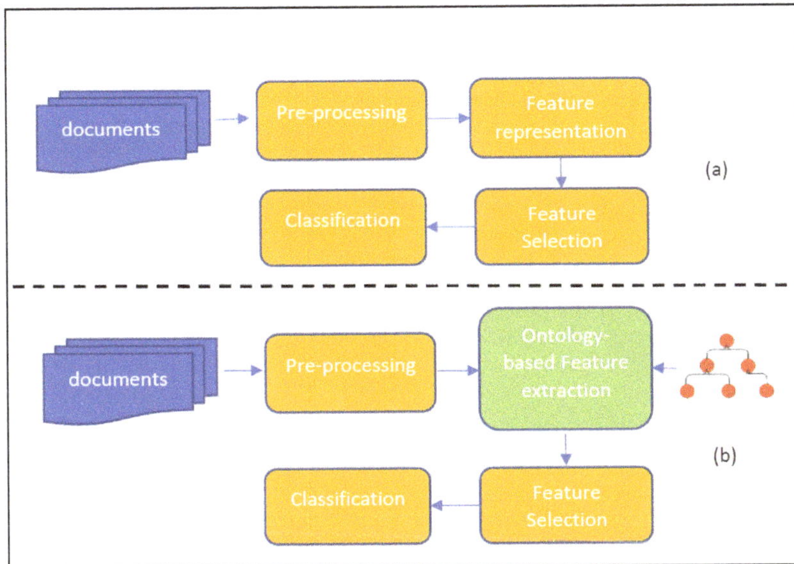

Figure 2. (**a**) Standard feature selection. (**b**) Ontology-based feature selection.

The main research domain where ontologies have been employed in terms of selecting specific features is document classification, where predefined categories are associated with free-text unstructured documents based on their content. The continuous increase of volumes of text documents on the Web makes text classification an important tool for searching information. Due to their enormous scale in terms of the number of classes, training examples, features, and feature dependencies, text classification applications present considerable research challenges.

In the following paragraphs we present related works organized according to selected and representative application domains. For each domain, we provide a summarized description of the related work and a table that organizes their main features according to specific criteria.

5.1. Document Classification

As presented in Table 1, there are several works related to ontology-based document classification, in different domains, using different approaches and ontologies. In the following paragraphs we provide insights to a selected representative set of those works.

Elhadad et al. [55] use the WordNet [56] lexical taxonomy (as an ontology) to classify Web text documents based on their semantic similarities. In the first phase, a number of filters are applied to each document to extract an initial vector of terms, called Bag of Words (BoW), which represent the document space. In particular, a Natural Language Processing Parser (NLPP) parses the text and extracts words in the form of tagged components (part of speech), such as verbs, nouns, adjectives, etc. Words that contain symbolic characters, non-English words, and words that can be found in pre-existing stopping word lists, are eliminated. Furthermore, in order to reduce redundancy, stemming algorithms are used to replace words with equivalent morphological forms, with their common root. In the second phase, all words in the initial BoW are examined for semantic similarities with categories in WordNet. Specifically, if a path exists in the WordNet taxonomy, from a word to a WordNet category via a common parent (hypernym), then the word is retained, otherwise it is discarded. Once the final set of terms has been selected, the feature vector for each document is generated by assigning a weight to each term. Authors use the Frequency-Inverse Document Frequency (*TFIDF*) statistical measurement, since it computes the importance of a term t, both in an individual document and in the whole training set. *TFIDF* is defined as:

$$TFIDF(t) = TF(t) \times IDF(t) \tag{1}$$

where

$$TF(t) = \frac{Number\ of\ occurances\ of\ term\ t}{Total\ number\ of\ terms\ in\ doc} \tag{2}$$

and

$$IDF(t) = log\frac{Total\ Number\ of\ docs}{Number\ of\ docs\ with\ term\ t} \tag{3}$$

Effectively, terms that appear frequently in a document, but rarely in the overall corpus, are assigned larger weights. Authors compared against the Principal Component Analysis (PCA) method and report superior classification results. However, they recognize that a limitation in their approach is that important terms that are not included in WordNet will be excluded from the feature selection.

Vicient et al. [57], employ the Web to support feature extraction from raw text documents, which describe an entity (symbolized with ae), according to a given ontology of interest. In the first step, the OpenNLP [58] parser analyzes the document and detects potential named entities (PNE) related to the ae, as noun phrases containing one or more words beginning with a capital letter. A modified Pointwise Mutual Information (PMI) measure is used to rank the PNE and identify those that are most relevant to the ae according to some threshold. In particular, for each $pne_i \in PNE$ probabilities are approximated by Web hit counts provided by a Web search engine,

$$NE_{score}(pne_i, ae) = \frac{WebHitsCount(pne_i \& ae)}{WebHitsCount(pne_i)} \tag{4}$$

In the second step, a set of Subsumer Concepts (SC) is extracted from the retained Named Entities (NE). To do so, the text is scanned for instances of certain linguistic patterns that contain each $ne_i \in NE$. Each pattern is used in a Web query and the resulting Web snippets determine the subsumer concepts representing the ne_i. Next, the extracted SC are mapped to ontological classes (OC) from the input ontology. Initially, for each ne_i all its potential subsumer concepts are directly matched to lexical-similar ontological classes. If no matches are found then WordNet is used to expand the SC and direct matching is repeated. Specifically, the parents (hypernyms) in the WordNet hierarchy of each subsumer concept sc_i are added to SC. In order to determine which parent concepts are mostly relevant to the named entity ne_i, a search engine is queried for common appearances of the ae and the ne_i. The returned Web snippets are used to determine which parent synsets of sc_i are mostly related to ne_i. Synsets in Wordnet are groupings of words from the same lexical category that are synonymous and express the same concept. Finally, a Web-based version of the PMI measure, defined as

$$SOC_{score}(soc_i, ne_i, ae) = \frac{WebHitsCount(soc_i \& ne_i \& ae)}{WebHitsCount(soc_i \& ae)} \tag{5}$$

is used to rank each of the extracted ontological classes (soc_i), related to a named entity. The soc_i with the highest score that exceeds a threshold is used as annotation. The authors tested their method in the Tourism domain. For the evaluation, they compared precision (ratio of correct feature to retrieved features) and recall (ratio of correct features to ideal features) against manually selected features from human experts. They report 70–75% precision and more than 50% accuracy and argue that such results considerably assist the annotation process of textual resources.

Wang et al. [59] reduce the dimensionality of the text classification problem by determining an optimal set of concepts to identify document context (semantics). First, the document terms are mapped to concepts derived from a domain-specific ontology. For each set of documents of the same class, the extracted concepts are organized in a concept hierarchy. A hill-climbing algorithm [60] is used to search the hierarchy and derive an optimal set of concepts that represents the document class. They apply their method to classification of medical documents and use the Unified Medical Language System (UMLS) [61] as the underlying domain-specific ontology. UMLS query API is used to map document terms to concepts and to derive the concept hierarchy. For the hill-climbing heuristic, a frequency measure is assigned to each leaf concept node. The weight of parent nodes is the sum of the children's weights. Based on the assigned weights, a distance measure between two documents is derived, and used to define the fitness function. Test documents undergo the same treatment and are classified based on the extracted optimal representative concepts. For their experiments the authors use a KNN classifier and report improved accuracy, but admit that an obvious limitation of their method is that it is only applicable in domains that have a fully developed ontology hierarchy.

Khan et al. [62] obtain document vectors defined in a vector space model. This is accomplished in terms of the following steps. First, after identifying all the words in the documents, they remove the stop-words from the word data base, creating a BoW. Next, a stemming algorithm is applied to assign each word to its respective root word. Phrase frequency is estimated using a Part of Speech (PoS) tagger. Next, they apply the Maximal Frequent Sequence (MFS) [63] to obtain the highly frequent terms. MFS is a sequence of words that is frequent in a collection of documents, while it is not related to any other sequence of the same kind [63]. The final set of features is selected by examining similarities with ontology-based categories in WordNet [56] and applying a wrapper approach. Using the *TFIDF* statistical measure weights are assigned to each term. Finally, the classifier is trained in terms of the naive Bayes algorithm.

Abdollali et al. [64] also address feature selection in the context of classification of medical documents. In particular, they aim at distinguishing clinical notes that reference Coronary Artery Disease (CAD) from those that do not. Similarly to [59], they use a query tool (MetaMap) to map meaningful expressions, in the training documents to concepts in UMLS. Since, their target is CAD documents, they only keep concepts such as "Disease or Syndrome" and "Sign or Symptom" and discard the rest. The retained concepts are assigned a *TFIDF* weight to form the feature vector matrix that will be used in the classification. In the second stage, the particle swarm optimization [65] algorithm is used to select the optimal feature subset. The particles are initialized randomly by numbers in $[-1, 1]$, where a positive number indicates an active feature while a negative value an inactive one. The fitness function for each particle is based on the classification accuracy,

$$Fitness(S) = \frac{TP + TN}{TP + TN + FP + FN} \tag{6}$$

where S represents the features set, TP (True Positive) and FP (False Positive) are the number of correctly and incorrectly identified documents and TN (True Negative) and FN (False Negative) the number of correctly and incorrectly rejected documents. The particle's fitness value is estimated as the average of the accuracies using a 10-fold cross validation procedure. The authors evaluated their method using five classifiers (NB, LSVM, KNN, DT, LR) and reported both significant reduction of the feature space and improved accuracy of the classification in most of their tests.

Lu et al. [66] attempt to predict the probability of hospital readmission within 30 days after a heart failure, by means of the medication list prescribed to patients during their initial hospitalization. In the first stage, the authors combine two publicly accessible drug ontologies, namely RxNorm [67] and NDF-RT [68], into a tree structure, that represents the hierarchical relationship between drugs. The RxNorm ontology serves as drug thesaurus, while NDF-RT as drug functionality knowledge base. The combined hierarchy consists of six class levels. The top three levels correspond to classes derived from the Legacy VA class list in NDF-RT and represent the therapeutic intention of drugs. The fourth level represents the general active ingredients of drugs. The fifth level refers to the dosage of drugs and uses a unique identifier to match drugs to the most representative class in RxNorm (RXCUI). The lowest level refers to the dose form of drugs and uses the local drug code used by different hospitals. Each clinical drug corresponds to a single VA class, a single group of ingredients, and a single RxNorm class. In the second stage, a top-down depth-first traversal of the tree hierarchy is used to select a subset of nodes as features. For each branch, the nodes are sorted according to the information gain ratio ($IGR(F) = IG(F)/H(F)$). The features in the ordered list are marked for selection one by one, while parent and child features with lower scores are removed from the list. In order to evaluate their method, the authors use the naive Bayes classifier and employ the area under the receiver operating characteristic curve to evaluate its performance. Their experiments showed that the ontology-guided feature selection outperformed the other non-ontology-based methods.

Barhamgi et al. [69] explore the use of the semantic web and domain ontologies to automatically detect indicators and warning signals, emitted from messages and posts in social networks, during the radicalization process of vulnerable individuals and their recruitment from terrorist organizations. Specifically they devise an ontology for the radicalization domain and exploit it to automatically annotate social messages (tweets). These annotations are combined with a reasoning mechanism to infer values of pre-determined radicalization indicators according to a set of inference rules. The ontology is built in two steps. First, a group of experts define and organize the main classes, properties and relationships. These are related to high level concepts, which represent the radicalization indicators, namely "Perception of discrimination for being Muslim", "Expressing negative ideas about Western society", "Expressing positive ideas about jihadism", "The individual is frustrated" and "The individual is introvert". In the second step, each concept or instance is expanded with a set of related keywords from the BabelNet knowledge base, which are inserted to the ontology as OWL annotation properties. Using the enriched ontology, the Ontology Classifier module annotates input messages with low-level concepts and the Ontology Instantiator module populates the ontology with the annotated messages and associated users. The Ontology Reasoner executes a set of SWRL rules on the populated domain ontology and infers new concepts for the messages, which are also added in ontology as new semantic annotations. Finally, the Ontology Querying module is used to compute the radicalization indicators for each user of the considered dataset by executing specific queries on the populated ontology. The proposed system was tested on a randomly selected dataset containing radical and neutral Twitter messages against a baseline using the standard F1 score. The obtained results showed that the ontology-based approach gave higher precision and recall and overall better classification results.

Kerem and Tunga [70] describe a framework to investigate the effectiveness of using WordNet semantic features in text categorization. In particular, they examine the effect that part-of-speech tagging, inclusion of WordNet features, and word sense disambiguation, have on text classification. POS features consist of the nouns, verbs, adjectives, and adverbs in the document. WordNet features are the synsets with specific relations to the document terms, namely synonyms, hypernyms, hyponyms, meronyms, and topics. Word sense disambiguation is performed in order to exclude irrelevant synsets from the feature set. For each synset, a score is computed as the sum of its similarities (common hypernyms and topics) to all other synsets. Synsets with a score below a predefined threshold are excluded. Experiments were performed using the SVM classifier on five standard datasets. The contribution of each task to the classification was quantified by means of the macro and micro variants of the F-measure metric. The authors conclude that using nouns, adjectives, and verbs in conjunction with the raw terms improve the classification, while adverbs have an adverse effect. Meronyms, hyponyms, topics, synonyms and hypernyms further improved the classification, in increasing order of importance. Finally, disambiguation was also found to benefit the classification.

Fodeh et al. [71] study the effect that incorporating ontology information in the feature selection process has on document clustering. First, they discuss a simple technique for feature selection and compare it against a word sense disambiguation process (WSD), where semantic relations have been considered in the document clustering. The simple procedure consists of a pre-processing step, which includes stop-word removal and stemming, and a cleaning step, where non-nouns are removed and only stemmed terms, identified as nouns are retained. Noun identification takes place with a simple lookup in the WordNet noun database. The WSD procedure replaces the selected nouns by their most appropriate senses (concepts) as used in the context of the document. Formally, if $\delta(s_q, s_p)$ denotes the similarity between two senses s_q and s_p, and $S_i = \{s_{i1}, s_{i2}, \ldots, s_{ik}\}$ is the set of all senses associated with noun t_i according WordNet, then its most appropriate sense $\hat{s}_i$ is the sense s_{il}, which maximizes the sum of maximum similarities with the senses of all the other terms in a document d,

$$\hat{s}_i = \arg\max_{s_{il} \in S_i} \sum_{t_j \in d} \max_{s_{jm} \in S_j} \delta(s_q, s_p) \tag{7}$$

The authors apply the Wu–Palmer similarity measure and restrict their consideration to the first three senses of each noun. The comparison showed that in most cases (12 out of the 19 tested datasets) WSD failed to justify its increased cost as it did not improve upon the results obtained by the simple approach. Next, the authors examine the effect of polysemous and synonymous nouns in clustering. Specifically, five types of feature sets are compared, namely all nouns X_{all}, all polysemous nouns X_{poly}, all synonymous nouns X_{syn}, the union $X_{both} = X_{poly} \cup X_{syn}$, and three random subsets of nouns $X_{rand} \subset X_{all} \setminus X_{both}$. For each feature set the pairwise document cosine similarity matrix is correlated to that obtained using X_{all}. Additionally, the purity of clusters obtained from each feature set are compared. The analysis showed that the correlation using polysemous and synonymous nouns is always high, indicating that those nouns strongly participate in the assembly of the final clusters and that their inclusion produces clusters with higher purity. In the final stage of their study, the authors build upon their previous conclusions and propose a feature selection framework for clustering, which utilizes a small subset of the semantic features extracted from WordNet, named core semantic features (CSF). In particular, a noun is considered a core feature if it is polysemous or synonymous and in the top 30% of the most frequent nouns. Furthermore, after disambiguation the noun should achieve either an information gain greater than a predefined threshold or zero entropy. The method was tested on several (19) datasets using spherical K-means clustering and the authors report at least 90% feature reduction while mainting and in some cases improving cluster purity, compared to using all nouns or concepts. However, due to the small number of CSF some documents may not include any of those features and thus be left uncovered. The authors solve this problem by applying a modified centroid mapping.

Garla and Brandt [72] propose two ontology-guided feature engineering methods, which utilize the UMLS Ontology for classification of clinical documents. The first method constitutes an ontology-guided feature ranking technique, based on an enhanced version of standard Information Gain (IG). The enhancement lies in the fact that the IG assigned to a feature c, considers also documents that do not directly contain it, but instead contain any children of feature c or its hypernyms in the UMLS hierarchy. This is referred to as the imputed information gain (IG_{imp}). Features with a value of IG_{imp} below a predefined threshold are discarded. In addition, the imputed IG is combined with the Lin [73] measure to construct a context-dependent semantic similarity kernel, referred to as supervised Lin measure. Given concepts c_1 and c_2 the Lin similarity measure is defined as

$$sim_{lin}(c_1, c_2) = \frac{2 \cdot IC(LCS(c_1, c_2))}{IC(c_1) + IC(c_2)}, \tag{8}$$

where $IC(c) = -log(freq(c))$ is the information content of concept c,

$$freq(c) = freq(c, C) + \sum_{c_s \in children(c)} freq(c_s), \tag{9}$$

the frequency of concept c in document C and $LCS(c_1, c_2)$ the least common subsumer of concepts c_1 and c_2. If $IG_{imp}(LCS(c_1, c_2))$ exceeds a predefined threshold then the Lin measure determines the similarity of c_1 and c_2, otherwise the similarity is set to 0. The two techniques were evaluated on a standard dataset using the SVM classifier. Performance was measured with the macro-averaged F1 score. The authors report that the results match those of other top systems. Both imputed information gain and supervised Lin measure improved the classification, however the latter only marginally. They recognize that one limitation of their study is the small corpus size used to compute the semantic similarity measures and intend to experiment with other similarity measures, which do not depend on the corpus size.

In [74], Qazia and Goudar study the effectiveness of ontology in the classification of Web documents. They are interested in a corpus with Web pages in four distinct categories from the domain of sports, namely cricket, football, hockey, and baseball. First, they use OWL to develop the ontology which represents the set of classes, individuals, and relationships for the domain under consideration. Then an ontology guided term-weighting technique is applied to extract and weight the feature terms that represent each document. Specifically, after stop-word removal and stemming, each obtained term from the Web page is looked-up in the Ontology. If the term is not found it is discarded, otherwise the sum of its TF-IDF scores from each document, is assigned as its semantic weight,

$$w_{term} = \sum_{doc_j \in Documents} |term\ in\ doc_j| \cdot log\frac{|Documents|}{|Documents_{term}|}, \tag{10}$$

where w_{term} is the term semantic weight and $Documents_{term}$ are the documents that contain the term. The authors compared their method against the standard Bag of Words approach with TF-IDF term weighting and report that the use of ontology considerably improved the performance of the classification.

Table 1. Document classification organized according to specific criteria.

Related Work	Application	Ontology	Feature Selection	Classifier
[55]	Web Text	WordNet	TFIDF	NB, JRip, J48, SVM
[56]	Web text	Tourism, Space, Film, WordNet	Web-based PMI (NE_{score}, SOC_{score})	-
[59]	Text (Medical)	UMLS	Frequency, Hill Climbing	KNN
[62]	Text	WordNet	MFS, TFIDF	NB
[64]	Clinical Notes (CAD)	UMLS	TFIDF, PSO	NP, LSVM, KNN, DT, LR
[66]	Medication list (hospital re-entry)	RxNorm, NDF-RT	IGR	NB
[69]	Web text (Tweets)	Custom, BabelNet	Ontology-based, SWRL	-
[70]	Text	WordNet	Similarity	SVM
[71]	Text (Clustering)	WordNet	Similarity	Sperical K-means
[72]	Text (Clinical)	UMLS	IG, Lin	SVM
[74]	Web text (Sports)	Custom	TFIDF	MNB, DT, KNN, Rocchio
[75]	Text	WordNet, OpenCyc, SUMO	Custom (Mapping Score)	SVM

In their work [75], Rujiang and Junhua attempt to improve text classification by replacing the traditional Bag of Words (BoW) document representation with Bag of Concepts (BoC) derived from multiple relevant ontologies. Initially, they employ the Jena Ontology API [76] to combine three ontologies, namely WordNet, OpenCyc, and SUMO. In order to align equivalent concepts, first they identify homographic concepts. Two concepts are homographic, when they belong to different ontologies, but share the same name or the same synonym. Homographic concepts are also equivalent, if their direct subconcepts or superconcepts, with respect to a particular relation type, are homographic. Once the ontologies have been aligned, a context is obtained for each concept c in the set of all ontologies ($Ocont(c)$), as the union of all synonyms of c, the names of all subconcepts and superconcepts of c and the synonyms of the subconcepts and superconcepts. In the next step, the documents are pre-processed, including tokenization, stop-word elimination, stemming, and part-of-speech tagging. For each word in a document's cleaned up word-list, a context is obtained ($Wcont(w)$) as the set of all stems for all words in the document. When the stem of a concept c or one of its synonyms matches a word w and the POS of c is the same as that of w, a mapping score is assigned to w, indicating how well it maps to c. This mapping score (ms) is defined as the ratio of the number of elements that occur in both word and ontology contexts to the number of elements of the latter,

$$ms(w,c) = \frac{|Wcont(w) \cap Ocont(c)|}{|Ocont(c)|} \tag{11}$$

The method was tested on the Neuters-21758, OSHUMED, and 20NG datasets with a linear SVM classifier. Performance was measured with the standard micro and macro F1 metrics. The authors conclude that the BoC representation improved the classification compared to BoW.

5.2. Opinion Mining

Opinion mining is also called sentiment analysis [77]. In general, the methodologies that deal with sentiment analysis focus on classifying a document as having a positive or negative polarity, regarding a pre-specified objective [78–81]. Certain difficulties in implementing the above strategy led to the necessity of using ontology-based features [80,82]. An example of such a difficulty is related to the fact that positive (negative) document on an object does not imply that the user has positive (negative) opinion regarding the whole set of features assigned to that document [80,81]. The ontology-based feature selection for sentiment analysis is a complex and difficult endeavor, mainly because it involves high semantic representations of expressed opinions along with diversified characteristics encoded in the ontology as well as in the corresponding features [78,80,83].

The general implementation framework of ontology-based feature selection for opinion mining is given in Figure 3. Three basic levels are identified.

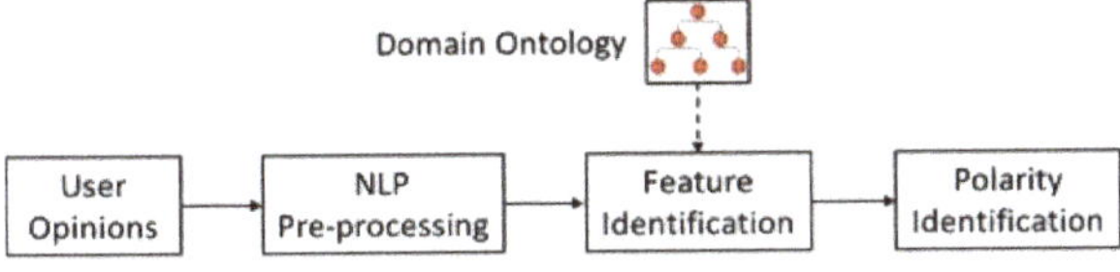

Figure 3. General algorithmic framework for ontology-based sentiment analysis.

The first level concerns the pre-processing of the users' opinions in terms of Natural Language Processing (NLP) techniques. The objective is to perform linguistic and syntactic process of the available textual data. This task can be accomplished by implementing several NLP tools such as stemming, tokenization, Part of Speech (PoS) tagging, morphological analysis, syntax parsing, etc. [81,84]. As a result of the NLP pre-processing, an initial set of features is extracted and fed into the next level for further processing. The second level carries out the implementation of a domain ontology to identify the most important features included in initial set of features. The domain ontology can be generic [80,83] or custom (i.e., created for a specific application) [77,82,83]. As seen in Figure 3, the input to this step is the pre-processed corpus of opinions as well as the domain ontology, while its output comes in the form of potential features identified from the text, which are then represented in some convenient form (e.g., vector-based representation, etc.), which will help their elaboration by the next levels [80,83]. As far as the type of ontology is concerned, it can be a crisp ontology i.e., a precise (binary) specification of a conceptualization [79,80,83–85] or ontology based on fuzzy set theory [86,87]. To further reduce the feature space dimensionality, we can apply standard feature selection methods (e.g., PCA, chi-square, information gain), or strategies involve the calculation of pairwise similarity measures between features or score values, which are assigned to the features indicating their importance [80,83]. The third level deals with the polarity identification. Two common strategies involved in this level are machine learning methods (e.g., SVM, clustering,) [77,83] and lexicon-based approaches (e.g., SentiWordNet, SentiLex, OpLexicon) [80,85]. The implementation of machine-learning techniques utilizes a set of training data and involves iterative classification processes. On the other hand, lexicon-based approaches rely on the application of batch procedures, and once they have been built, no training data are necessary.

Table 2 illustrates the basic characteristics of various approaches that exist in the literature. In the following paragraphs, we briefly describe those approaches.

Table 2. Related works organized according to specific criteria.

Related Work	Ontology	Type of Ontology	Classifier
[80]	Movie Ontology	Crisp	Lexicon-based
[83]	Movie Ontology, WordNet	Crisp	SVM
[77]	Custom based on FCA and OWL	Crisp	SVM
[85]	Movie Ontology	Crisp	Lexicon-based
[86]	Custom based on fuzzy set theory	Fuzzy	SVM
[82]	Custom	Crisp	Lexicon-based
[86]	Custom	Fuzzy	Lexicon-based
[84]	Custom	Crisp	Lexicon-based

Penalver-Martinez et al. [80] perform the feature identification by employing the Movie Ontology [88]. To further reduce the feature space dimensionality, they assign to each feature a score function of importance, which considers the position of the linguistic expression of a feature within the text. The score function is structured by separating the text in three disjoint parts namely, (a) the beginning, (b) the middle, and (c) the end. Given a feature f_i and a user's opinion text t_j, the above three text parts are symbolized as O_{aj}, O_{bj}, and O_{cj}. The number of occurrences of f_i in those three text parts are defined as $|O_{aj}|_i$, $|O_{bj}|_i$, and $|O_{cj}|_i$. By defining the respective importance degrees of occurrence of f_i in the above three text parts as z_{aj}^i, z_{bj}^i, and z_{cj}^i, the resulting score function reads as follows,

$$score(f_i, t_j) = z_{aj}^i|O_{aj}|_i + z_{bj}^i|O_{bj}|_i + z_{cj}^i|O_{cj}|_i \qquad (12)$$

The features are grouped in accordance with score value and attached to a main concept of the ontology. The set of the above concepts constitute the final set of features.

Finally, the polarity identification is carried out in terms of the SentiWordNet framework, where three possible outcomes are obtained namely, positive, neutral, and negative opinion.

Siddiqui et al. [83] used standard NLP techniques to identify several potential features, which are added onto a feature vector. Then, a semantic processing approach takes place to select the most important features. The semantic processing is conducted in a sequence of steps. The first step applies a lexical pruning algorithm aiming to discard all features that are not part of lexical categories of the WordNet ontology. Second, they combined the Movie Ontology [88] and the standard WordNet ontology and developed a semantic similarity-based approach, that consists of three pairwise similarity measures namely the semantic similarity measure, the semantic relatedness measure, and semantic distance measure. Third, the calculated values of the above-mentioned pairwise similarity measures are gathered in a table that reports all the possible pairs. This table assists the computation of the weights of importance of each feature in relation to the rest of the features. To this end, an iterative algorithm is developed, which gradually refines the initial set of features, until the most important features are identified. The above algorithm is based on defining an appropriate threshold value, and the importance of a feature is decided according to whether the respective overall weight of importance is greater than the above threshold or not. Finally, the polarity identification is carried out in terms of a binary classification approach (i.e., positive or negative polarity) using a support vector machine algorithm.

Shein and Nyunt [77] developed an ontology in OWL using formal concept analysis (FCA). The algorithmic framework attempts to form semantic structures that are formal abstraction of linguistic concepts and moreover to identify conceptual structures among data. The result is an ontological framework able to effectively analyze complex text structures and to reveal dependencies within the data. The polarity identification is carried out in terms of a support vector machine algorithm that performs binary feature classification, which can correspond to positive or negative polarity.

de Freitas and Vieira [85] have developed an opinion mining framework for the Portuguese language. The feature identification and selection are conducted by using the Movie Ontology [88], while the polarity identification takes place in terms of Portuguese opinion lexicons.

Andrea and Fabrizi [89] propose a novel method for sentiment classification of text documents. In particular, they determine the orientation of a term based on the classification of its glosses and its definitions in online dictionaries. In the training phase, a seed term set, representative of the two categories Positive and Negative, is provided as input. By means of a thesaurus (online dictionary) the set is expanded with additional terms, that are lexically related (synonymous) to the seed terms and, therefore, can also be considered as representative of the two categories. This process is applied iteratively, until no new terms are added. For each term in the final set a textual representation is constructed, by collating its glosses, as found in machine-readable dictionaries. In that sense the method is semi-supervised since except for the initial seed terms, all features are selected algorithmically, rather than by human experts. For both the expansion and gloss retrieval operations, the authors employed the Wordnet Ontology, mainly because of its ease of use for automatic processing. Moreover, glosses in WordNet have a regular format that allows the production of clean textual representations without the need for manual text cleaning. After removing stop words, each such representation is mapped to a numerical vector by the standard normalized TFIDF score of its terms. Finally, the set of all vectors is used to train a binary classifier. In the experiments three types of classifiers were used, namely the multinomial naive Bayes, support vector machines with linear kernels, and the PrTFIDF probabilistic version of the Rocchio learner [90]. The algorithm was shown to outperform other state-of-the-art methods in standard benchmarks, while also being computationally much less intensive.

5.3. Other Applications

In this section, we review a number of other interesting applications that integrate ontology-based feature selection methods. Specifically, the analysis concerns manufacturing processes, recommendation systems, urban management, and information security. Table 3 summarizes the basic characteristics of these approaches.

Table 3. Related works organized according to specific criteria.

Related Work	Application	Ontology	Feature Selection	Classifier
[91]	Recommender system	Custom domain ontology	Ontology-based	KNN summary
[92]	Recommender system	DBpedia	Information	KNN Gain
[93]	Recommender system	Movie Ontology	Filtering	Clustering
[94]	Manufacturing	Custom domain ontology	Similarity measure	Rule-based
[95]	Manufacturing	Custom domain ontology	Filtering	Rule-based
[96]	Manufacturing	Custom domain ontology	Filtering	Rule-based
[97]	Manufacturing	Custom domain ontology	Filtering	Rule-based (Pearson Coef.)
[98]	Manufacturing	Custom domain ontology	Filtering	Rule-based
[99]	Urban management	Custom domain ontology	Filtering	Random Forest
[100]	Information security	Custom domain ontology	Filtering	Decision Tree

An important application of ontology-based feature selection algorithms is the selection of manufacturing processes. Mabkhot et al. [94] describe an ontology-based Decision Support System (DSS), which aims at assisting the selection of a Suitable Manufacturing Process (MPS) for a new product. In essence, selected aspects of MPS are mapped to ontological concepts, which serve as features in rules used for case-based reasoning. Traditionally, MPS has relied on expert human knowledge to achieve the optimal matching between material characteristics, design specifications, and process capabilities. However, due to the continuous evolution in material and manufacturing technologies and the increasing product complexity, this task becomes more and more challenging for humans. The proposed DSS consists of two components, namely the ontology and the Case-based Reasoning Subsystem (CBR). The purpose of the ontology is to encode all the knowledge related to manufacturing in a way which enables the reasoner to make a recommendation for a new product design. It consists of three main concepts, the Manufacturing Process (MfgProcess), the Material (EngMaterial) and the Product (EngProduct). The MfgProcess concept captures the knowledge about manufacturing in subconcepts, such as casting, molding, forming, machining, joining, and rapid manufacturing. The properties of each manufacturing process are expressed in terms of shape generation capabilities, which describe the product shape features a process can produce, and range capabilities, which express the product attributes that can be met by the process such as dimensions, weight, quantity, and material. The EngMaterial concept captures knowledge about materials, in terms of material type (e.g., metal, ceramic) and material process capability (e.g., sand casting materials). The EngProduct concept encodes knowledge about products, defined in the form of shape features and attributes. The ontology facilitates the construction of rules, which associate manufacturing processes with engineering products, through the matching of appropriate features and attributes with main process characteristics and capabilities. The semantic Web rule language (SWRL [101]) has been used as an effective method to represent causal relations. The purpose of the CBR subsystem is to find the optimal product-to-process matching. It does so in two steps. First, it scans the ontology for a similar product. To quantify product similarity, appropriate feature and attribute similarity measures have been developed and human experts have been employed to assign proper weights to features and attributes. If a matching product is found then the corresponding process is presented to the decision maker, otherwise SWRL rule-based reasoning is used to find a suitable manufacturing process. Finally, the ontology is updated with the newly extracted knowledge. The authors presented a use case to demonstrate the usability and effectiveness of the proposed DSS and argue that in the future such systems will become more and more relevant.

In [96], Kang et al. develop an ontology-based representation model to select appropriate machining processes as well as the corresponding inference rules. The ontology is quantified in terms of features, process capability with relevant properties, machining process, and relationships between concepts. A reasoning inference mechanism is applied to obtain the final set of processes for individual features. The determination of the process that corresponds to the highest contribution is carried out through a solid mechanism that associates the capability of the candidate processes with the accuracy requirements of a specific feature. The appropriate machining process is, then, selected so that the relationship constraint between a pre-specified set of processes is met. The whole process selection scheme is neutral (i.e., general enough) in the sense that it does not depend on a specific restriction, and thus it constitutes a reusable platform.

Han et al. [97] also apply ontology within the mechanical engineering domain, in particular the field of Noise, Vibration, and Harshness (NVH). Similar to the previous work, authors map important aspects of noise identification to ontological concepts, which serve as features for reasoning. They propose an ontology-based system for identifying noise sources in agricultural machines. At the same time, their method provides an extensible framework for sharing knowledge for noise diagnosis. Essentially, they seek to encode prior knowledge relating noise sound signals (targets) with vibrational sound signals (sources) in an ontology, equipped with rules, and perform reasoning to identify noise sources based on the characteristics of test input and output sound signals (parotic noise). In order to build the ontology, first, professional experience, literature, and standard specifications were surveyed to extract the concepts related to NVH. The Protégé tool was used to convert the concept knowledge into an OWL ontology and implement the SWRL rules, which match sound source and parotic noise signals. The Pellet tool is employed for reasoning. To quantify the signal correlations, the time signals are converted to the frequency domain and the values for seven common signal characteristics are calculated. Specifically, relation of the frequency of the parotic signal to the ignition frequency, peak frequency, Pearson coefficient, frequency doubling, loudness, sharpness, and roughness. The effectiveness of the method was demonstrated in a use case, where the prototype system correctly identified the main noise source. After improving the designated area the noise was significantly reduced. The authors argue that the continuous improvement in the knowledge base and rule set of the ontology model has the potential to allow the design system to perform reasoning that simulates the thinking process of the expert in the field of NVH.

Belgiu et al. [99] develop an ontological framework to classify buildings based on data acquired with Airborne Laser Scanning (ALS). They followed five steps. Initially, they preprocessed the ALS point cloud and applied the eCognition software to convert it to raster data, which were used to delineate buildings and remove irrelevant objects. Additionally, they obtained values for 13 building features grouped in four categories: extent features, which define the size of the building objects, shape features, which describe the complexity of building boundaries, height, and slope of the buildings' roof. In the next step, human expert knowledge and knowledge available in literature were employed to define three general purpose building ontology classes, independent of the application and the data at hand, namely Residential/Small Buildings, Apartment/Block Buildings, and Industrial and Factory Buildings. In order to identify the metrics that were mostly relevant to the identification of building types, a set with 45 samples was used to train a random forest classifier with 500 trees and $\sqrt{m}$ features (m number of input features). The feature selection process identified slope, height, area, and asymmetry as the most important features. The first three were modeled in the ontology with empirically determined thresholds by the RF classifier. Finally, building type classification was carried out based on the formed ontology. The classification accuracy was assessed by means of precision, recall, and *F*-measure and the authors reported convincing results for class A while classes B and C had less accurate results. However, they argue that their method can prove useful for classifying building types in urban areas.

Finally, two interesting applications of ontology-based feature selection algorithms concern the Recommendation Systems (RS) and the information security/privacy research areas. In [91], Di Noia et al. develop a filter-based feature selection algorithm by incorporating ontology-driven data summarization for Linked Data (LD)-based Recommendation System (RS). The selection mechanism determines the k most important features in terms of the similarity between instances included in a given class of data summaries, which are generated by an ontology-based framework. Two types of descriptors are employed: pattern frequency and cardinality descriptors. A pattern is defined as a schema using an RDF triple denoted as (C, P, D), where C and D are classes or datatypes, and P is a property that expresses their relationship. C is called the source type and D the target type. The patterns are used to generate data summarization from a knowledge graph-based framework. Each pattern is associated with a frequency that corresponds to the number of relational assertions from which the pattern has been extracted. Therefore, a pattern frequency descriptor can be viewed as a set of statistical measures. A cardinality descriptor encodes information about the semantics of properties as used within specific patterns and can be used in computing the similarities between these patterns. To obtain the cardinality descriptors, the authors extended the above-mentioned knowledge graph framework. The LD and one or more ontologies are the inputs to the knowledge graph framework, while its outputs are: a type graph, a set of patterns along with the respective frequencies, and the cardinality descriptors. To this end, the filtering-based feature selection consists of two main steps. First, the cardinality descriptors are implemented to filter out features (i.e., pattern properties) that correspond to properties connecting one target type with many source types. Second, the pattern frequency descriptors are applied to rank in a frequency-based descending order all features and select the top-k features.

In [100], Guan et al. studied the problem of mapping Security Requirements (SR) to Security Patterns (SP). Viewing the SPs as features, feature selection is set up to perform the above mapping procedure. This selection is based on developing an ontology-based framework and a classification scheme. To accomplish this task, they described the SRs using four attributes namely, Asset (A), Threat (T), Security Attribute (SA), and Priority (P). The SRs are represented as rows in a two-dimensional matrix, where the columns correspond to the above attributes. Then, the meaning of each SR is: for a given asset A, one or more threats Ts may threat A by violating one or more attribute values of SA. In addition, each SR is to be fulfilled in a sequence according to the value of P during software development. Then, they generate complete and consistent SRs by eliciting values for the above attributes using the risk-based analysis proposed in [102]. On the other hand, Security Patterns (SP) are described in terms of three attributes namely, Context that defines the conditions and situation in which the pattern is applicable, Problem that defines the vulnerable aspect of an asset, and Solution that defines the scheme that solves the security. To intertwine the above information they developed a two-level ontological framework using an OWL-based security ontology. The first level concerns the ontology-based description of SRs and the second the ontology-based description of SPs. These descriptions were carried out by quantifying mainly the risk relevant and annotating security related information. To this end, a classification scheme selects an appropriate set of SPs for each SR. The classification scheme is developed by considering multiple aspects such as life-cycle, architectural layer that organizes information from low to high abstraction level, application concept that partitions the security patterns according to which part of the system they are trying to protect, and threat type that uses the security problems solved by the patterns.

5.4. Discussion

Based on the analysis of the related works, as organized in the previous subsections and the related tables, a number of findings can be summarized in the following lines:

a.   Most of the related works examined in this review paper concern ontology-based feature selection for text document classification, with the majority of them being Web-related.
b.   Most of the approaches utilize generic lexicons (either just the lexicon or in combination with domain ontologies), with the majority of them utilizing WordNet.
c.   For the task of feature selection, most of the approaches are based on the TFIDF method and filtering.
d.   SVM is the most common classification method, with KNN to follow.

6. Open Issues and Challenges

Features show dependencies among each other and, therefore, they can be structured as trees or graphs. Ontology-based feature selection in the era of knowledge graphs such as Wikidata, DBpedia, Freebase, and YAGO, can be influenced by two issues [103]:

a.   The large expansion of knowledge recorded in Wikipedia, from which DBpedia and YAGO have been created as reference sources for general domain knowledge, is needed to assist information disambiguation and extraction.
b.   Advancements in statistical NLP techniques, and the appearance of new techniques that combine statistical and linguistic ones.

An important and open issue in this domain is the linking of one document-mentioned entity to a particular KG's entity and the way it affects how other surrounding document entities are linked. Furthermore, it is more and more common nowadays to see an increasing number of inter-task dependencies being modeled, where pairs of tasks such as Named-Entity Recognition (NER) and Entity Extraction and Linking (EEL), Word Sense Disambiguation (WSD) and EEL, or EEL and Relation Extraction and Linking (REL), are seen as interdependent. The combinatorial approach of those tasks will continue to exist and advance since it has been proven highly effective to the precision of the overall information/knowledge extraction process. Regarding the contributed communities in this area of research, related works have been conducted by the Semantic Web community as well as from others such as the NLP, AI, and DB communities. Works conducted by the NLP community focus more on unstructured input, while database and data-mining-related works target more to semi-structured input [103].

As mentioned above, ontologies play a key role in feature selection. However, the engineering of ontologies, despite advancing quickly over the last decade, has not yet reached the status were consensus in domain-specific communities will deliver gold-standard ontologies for each case and application area. On the other hand, several issues and challenges related to the collaborative engineering of reused and live ontologies have been recently reported [50], indicating that this topic is still active and emerging. For instance, as far as concerns feature selection, different ontologies of the same domain used in the same knowledge extraction tasks will most probably result in a different set of features selected (schema-bias). Furthermore, human bias in conceptualizing context during the process of engineering ontologies (in a top-down collaborative ontology engineering approach) will inevitably influence the feature selection tasks. Specifically, in the cases where large KGs (e.g., DBpedia) are used for knowledge extraction, such a bias is present in both conceptual/schema (ontology) and data (entities) levels. Debiasing KGs is a key challenge in the Semantic Web and KG community itself [104], and consequently in the domain of KG-based feature selection.

Important challenges arise when ontology-based feature selection is applied to Linked Data (LD). LD appears to be one of the main structural elements of Big Data. For example, data created in social media platforms are mainly LD. LD appear to have significant correlations regarding various types of links and therefore, they possess more complex structure than the traditional attribute-valued data. However, they provide extra, yet valuable, information [105]. The challenges of using ontology-based feature selection in LD concern the development of ontology-based frameworks to exploit complex relation between data samples and features, and how to use them in performing effective feature selection, and to evaluate the relevance of features without the guide of label information.

Another interesting research area is the real-time feature selection. The main difficulty in dealing with real-time feature selection is that both data samples and new features must be taken into account simultaneously. Most of the methods that exist in the literature rely on feature pre-selection or on feature selection without online classification [106,107]. On the other hand ontologies encoded in trees or knowledge graphs may provide some benefits such as solid representations of the current relations between features, which can be used to predict any possible relation between the current available features and the ones that are expected to arise in real-time processing tasks. Therefore, to develop ontology-based feature selection methods for achieving real-time analysis and prediction regarding high-dimensional datasets remains a challenge.

Finally, an important open issue to consider is scalability. Scalability quantifies the impact imposed by increasing the training data size on the computational performance of an algorithm in terms of accuracy and memory [105,107]. The basics of feature selection and classification were developed before the era of Big Data. Therefore, most feature selection algorithms are not efficient in scaling high-dimensional data as their efficiency appears to reduce quickly. On the other hand, scaling-up favors the accuracy of the model. Therefore, there is a trade-off between finding an appropriate set of features and the model's accuracy. In this direction, the challenge is to define appropriate ontology-based relations between features in order to group them in such a way that the resulting set of features will be able to maintain acceptable model's accuracy.

7. Conclusions

This study provided an overview of ontology-based feature selection for classification processing. The presented approaches in selected application domains showed that ontologies can effectively uncover dominant features in diverse knowledge domains and can be integrated into existing feature selection and classification algorithms. Specifically, in the context of text classification, domain-specific ontologies combined mainly with the Word-Net taxonomy, can be utilized to map terms in documents to concepts in the ontology, thus replacing specific term-based document features with abstract and generic concept-based features. The latter capture the content of the text and can be used to train accurate and efficient classifiers. In the field of manufacturing engineering, ontologies can be employed to map human knowledge to concepts that serve as features for case-based reasoning and support decision making, such as selection of manufacturing process or noise source identification. In the domain of urban management, building type recognition can be facilitated by ontology. Moreover, the benefits of using an ontology-based framework to drive feature selection were investigated regarding software development/engineering applications such as recommendations systems and security information/privacy approaches. Finally, certain open issues and challenges were discussed and a number of relevant problems were identified. Although, this survey is by no means exhaustive, it demonstrates the broad applicability and feasibility of ontology-based feature extraction and selection.

Author Contributions: K.S. investigated and wrote the classification, feature selection problem, and wrote the first draft of the paper; G.E.T. proposed the idea, and investigated the ontology-based feature problem; K.K. investigated the ontology-based frameworks and wrote the respective sections; All authors contributed to the final version of the paper. All authors have read and agreed to the published version of the manuscript.

Funding: This research received no external funding.

Institutional Review Board Statement: Not applicable.

Informed Consent Statement: Not applicable.

Data Availability Statement: Not Applicable, the study does not report any data.

Acknowledgments: The authors want to thank the reviewers for their effort to provide their valuable comments.

Conflicts of Interest: The authors declare no conflict of interest.

Abbreviations

The following abbreviations are used in this manuscript:

ALS	Airborne Laser Scanning
BoC	Bag of Concepts
BoW	Bag of Words
CAD	Coronary Artery Disease
CARs	Class Association Rules
CBR	Case-based reasoning subsystem
DSS	Decision support system
DC	Dublin Core
DT	Decision Tree
EEL	Entity Extraction and Linking
FE	Feature Engineering
FOAF	Friend Of A Friend
IG	Information Gain
IGR	Information Gain Ratio
KNN	K-Nearest neighbor
LCS	Least Common Subsumer
LD	Linked data
LR	Logistic Regression
LSVM	Linear Support Vector Machine
LVQ	Learning Vector Quantization
MFS	Maximal frequent sequence
MPS	Suitable manufacturing process
NB	Naive Bayes
NDF-RT	National Drug File - Reference Terminology
NE	Named entities
NER	Named-Entity Recognition
NLPP	Natural Language Processing Parser
NVH	Noise, Vibration and Harshness
OC	Ontological classes
OWL	Web Ontology Language
PCA	Principal Component Analysis
PMI	Pointwise Mutual Information
PoS	Part of speech
PSO	Particle Swarm Optimization
RBC	Rule-Based Classification
RDF	Resource Description Framework
REL	Relation Extraction and Linking
RDFS	RDF Scheme
RF	Random Forest
RS	Recommendation (or Recommender) systems
SC	Subsumer concept
SOM	Self-organizing Map

SP	Security patterns
SR	Security requirements
SVM	Support Vector Machines
SWRL	SemanticWeb rule language
TFIFD	Term frequency-inverse document frequency
UMLS	Unified Medical Language System
XML	eXtensible Markup Language
WSD	Word Sense Disambiguation

References

1. Heilman, C.M.; Kaefer, F.; Ramenofsky, S.D. Determining the appropriate amount of data for classifying consumers for direct marketing purposes. *J. Interact. Mark.* **2003**, *17*, 5–28. [CrossRef]
2. Kuhl, N.; Muhlthaler, M.; Goutier, M. Supporting customer-oriented marketing with artificial intelligence: Automatically quantifying customer needs from social media. *Electron. Mark.* **2020**, *30*, 351–367. [CrossRef]
3. Kour, H.; Manhas, J.; Sharma, V. Usage and implementation of neuro-fuzzy systems for classification and prediction in the diagnosis of different types of medical disorders: A decade review. *Artif. Intell. Rev.* **2020**, *53*, 4651–4706. [CrossRef]
4. Tomczak, J.M.; Zieba, M. Probabilistic combination of classification rules and its application to medical diagnosis. *Mach. Learn.* **2015**, *101*, 105–135. [CrossRef]
5. Kumar, A.; Sinha, N.; Bhardwaj, A. A novel fitness function in genetic programming for medical data classification. *J. Biomed. Inform.* **2020**, *112*, 103623. [CrossRef] [PubMed]
6. Jiménez-Guarneros, M.; Gómez-Gil, P. Standardization-refinement domain adaptation method for cross-subject EEG-based classification in imagined speech recognition. *Pattern Recognit. Lett.* **2021**, *141*, 54–60. [CrossRef]
7. Langari, S.; Marvi, H.; Zahedi, M. Efficient speech emotion recognition using modified feature extraction. *Inform. Med. Unlocked* **2020**, *20*, 100424. [CrossRef]
8. Shah Fahad, M.; Ranjan, A.; Yadav, J.; Deepak, A. A survey of speech emotion recognition in natural environment. *Digit. Signal Process.* **2021**, *110*, 102951. [CrossRef]
9. Memon, J.; Sami, M.; Khan, R.A.; Uddin, M. Handwritten Optical Character Recognition (OCR): A Comprehensive Systematic Literature Review (SLR). *IEEE Access* **2020**, *8*, 142642–142668. [CrossRef]
10. Ma, Z.; Nie, F.; Yang, Y.; Uijlings, J.R.R.; Sebe, N.; Hauptmann, A.G. Discriminating Joint Feature Analysis for Multimedia Data Understanding. *IEEE Trans. Multimed.* **2012**, *14*, 1662–1672. [CrossRef]
11. Yang, Y.; Ma, Z.; Hauptmann, A.G.; Sebe, N. Feature Selection for Multimedia Analysis by Sharing Information Among Multiple Tasks. *IEEE Trans. Multimed.* **2013**, *15*, 661–669. [CrossRef]
12. Pashaei, E.; Aydin, E.N. Binary black hole algorithm for feature selection and classification on biological data. *Appl. Soft Comput.* **2017**, *56*, 94–106. [CrossRef]
13. Kim, K.; Zzang, S.Y. Trigonometric comparison measure: A feature selection method for text categorization. *Data Knowl. Eng.* **2019**, *119*, 1–21. [CrossRef]
14. Lee, Y.-H.; Hu, P.J.-H.; Tsao, W.-J.; Li, L. Use of a domain-specific ontology to support automated document categorization at the concept level: Method development and evaluation. *Expert Syst. Appl.* **2021**, *174*, 114681. [CrossRef]
15. Rezaeipanah, A.; Ahmadi, G.; Matoori, S.S. A classifcation approach to link prediction in multiplex online ego social networks. *Soc. Netw. Anal. Min.* **2020**, *10*, 27. [CrossRef]
16. Selvalakshmi, B.; Subramaniam, M. Intelligent ontology based semantic information retrieval using feature selection and classification. *Clust. Comput.* **2019**, *22*, S12871–S12881. [CrossRef]
17. Alzamil, Z.; Appellbaum, D.; Nehmer, R. An ontological artifact for classifying social media: Text mining analysis for financial data. *Int. J. Account. Inf. Syst.* **2020**, *38*, 100469. [CrossRef]
18. Everitt, B.S.; Landau, S.; Leese, M.; Stahl, D. *Cluster Analysis*; John Wiley and Sons: West Sussex, UK, 2011.
19. Wierzchon, S.T.; Klopotek, M.A. *Modern Algorithms of Cluster Analysis*; Springer: Berlin/Heidelberg, Germany, 2018.
20. Lyu, S.; Tian, X.; Li, Y.; Jiang, B; Chen, H. Multiclass Probabilistic Classification Vector Machine. *IEEE Trans. Neural Netw. Learn. Syst.* **2020**, *31*, 3906–3919. [CrossRef]
21. Shahrokni, A.; Drummond, T.; Fleuret, F.; Fua, P. Classification-Based Probabilistic Modeling of Texture Transition for Fast Line Search Tracking and Delineation. *IEEE Trans. Pattern Anal. Mach. Intell.* **2009**, *31*, 570–576. [CrossRef]
22. Demirkus, M.; Precup, D.; Clark, J.J.; Arbel, T. Hierarchical Spatio-Temporal Probabilistic Graphical Model with Multiple Feature Fusion for Binary Facial Attribute Classification in Real-World Face Videos. *IEEE Trans. Pattern Anal. Mach. Intell.* **2016**, *38*, 1185–1203. [CrossRef] [PubMed]
23. Zhou, H.F.; Zhang, J.W.; Zhou, Y.Q.; Guo, X.J.; Ma, Y.M. A feature selection algorithm of decision tree based on feature weight. *Expert Syst. Appl.* **2021**, *164*, 113842. [CrossRef]
24. Rincy, T.; Gupt, R. An efficient feature subset selection approach for machine learning. *Multimed. Tools Appl.* **2021**, *80*, 12737–12830.
25. Lu, X.-Y.; Chen, M.-S.; Wu, J.-L.; Chang, P.-C.; Chen, M.-H. A novel ensemble decision tree based on under-sampling and clonal selection for web spam detection. *Pattern Anal. Appl.* **2018**, *21*, 741–754. [CrossRef]

26. Gupta, K.; Khajuria, A.; Chatterjee, N.; Joshi, P.; Joshi, D. Rule based classification of neurodegenerative diseases using data driven gait features. *Health Technol.* **2019**, *9*, 547–560. [CrossRef]
27. Verikas, A.; Guzaitis, J.; Gelzinis, A.; Bacauskiene, M. A general framework for designing a fuzzy rule-based classifier. *Knowl. Inf. Syst.* **2011**, *29*, 203–221. [CrossRef]
28. Almaghrabi, F.; Xu, D.-L.; Yang, J.-B. An evidential reasoning rule-based feature selection for improving trauma outcome prediction. *Appl. Soft Comput.* **2021**, *103*, 107112. [CrossRef]
29. Singh, N.; Singh, P.; Bhagat, D. A rule extraction approach from support vector machines for diagnosing hypertension among diabetics. *Expert Syst. Appl.* **2019**, *130*, 188–205. [CrossRef]
30. Liu, M.-Z.; Shao, Y.-H.; Li, C.-N.; Chen, W.-J. Smooth pinball loss nonparallel support vector machine for robust classification. *Appl. Soft Comput.* **2021**, *98*, 106840. [CrossRef]
31. Aggarwal, C.C. *Data Classification: Algorithms and Applications*; CRC Press: Boca Raton, FL, USA, 2014.
32. Bishop, C.M. *Pattern Recognition and Machine Learning*; Springer: Singapore, 2006.
33. Verikas, A.; Gelzinis, A.; Bacauskiene, M. Mining data with random forests: A survey and results of new tests. *Pattern Recognit.* **2011**, *44*, 330–349. [CrossRef]
34. Padillo, F.; Luna, J.M.; Ventura, S. LAC: Library for associative classification. *Knowl. Based Syst.* **2020**, *193*, 105432. [CrossRef]
35. Deng, N.; Tian, Y.; Zhang, C. *Support Vector Machines: Optimization Based Methods, Algorithms, and Extensions*; Chapman and Hall/CRC: Boca Raton, FL, USA, 2013.
36. Nocedal, J.; Wright, S.J. *Numerical Optimization*; Springer: Berlin/Heidelberg, Germany, 2006.
37. Aha, D.W.; Kibler, D.; Albert, M.K. Instance-based learning algorithms. *Mach. Learn.* **1991**, *6*, 37–66. [CrossRef]
38. Mitchell, T. *Machine Learning*; McGraw-Hill; New York, NY, USA, 1997.
39. Duboue, P. *The Art of Feature Engineering: Essentials for Machine Learning*; Cambridge University Press: Cambridge, UK, 2020.
40. Liu, H.; Motoda, H. *Computational Methods of Feature Selection*; Chapman and Hall/CRC: Boca Raton, FL, USA, 2007.
41. Kuhn M.; Johnson, K. *Feature Engineering and Selection: A Practical Approach for Predictive Models*; Chapman and Hall/CRC Press: Boca Raton, FL, USA, 2020.
42. Guyon, I.; Elisseeff, A. An introduction to variable and feature selection. *J. Mach. Learn. Res.* **2003**, *3*, 1157–1182.
43. Jovic, A.; Brkic, K.; Bogunovic, N. A review of feature selection methods with applications. In Proceedings of the 38th International Convention on Information and Communication Technology, Electronics and Microelectronics (MIPRO), Opatija, Croatia, 25–29 May 2015; pp. 1200–1205.
44. W3C. OWL Use Cases and Requirements. 2004. Available online: https://www.w3.org/TR/2004/REC-webont-req-20040210/ (accessed on 16 June 2021).
45. OWL Reference. 2004. Available online: https://www.w3.org/OWL/ (accessed on 16 June 2021).
46. Dublin Core Metadata Initiative. 2000. Available online: https://dublincore.org/ (accessed on 16 June 2021).
47. Dan Brickley and Libby Miller. FOAF Vocabulary Specification 0.99. 2001. Available online: http://xmlns.com/foaf/spec/ (accessed on 16 June 2021).
48. The Gene Ontology Resource. 2008. Available online: http://geneontology.org/ (accessed on 16 June 2021).
49. Schema.org. Available online: http://schema.org/ (accessed on 16 June 2021).
50. Kotis, K.; Vouros, G.A.; Spiliotopoulos, D. Ontology engineering methodologies for the evolution of living and reused ontologies: Status, Trends, Findings and Recommendations. *Knowl. Eng. Rev.* **2020**, *35*, e4. [CrossRef]
51. Allemang, D.; Hendler, J. *Semantic Web for the Working Ontologist: Effective Modeling in RDFS and OWL*; Morgan Kaufmann Publishers Inc.: San Francisco, CA, USA, 2011.
52. Antoniou, G.; Groth, P.; van Harmelen, F.; Hoekstra, R. *A Semantic Web Primer*; The MIT Press: Cambridge, MA, USA, 2012.
53. Domingue, J.; Fensel, D.; Hendler, J.A. *Handbook of Semantic Web Technologies*; Springer: Heidelberg, Germany, 2011.
54. Tosi, D.; Morasca, S. Supporting the semi-automatic semantic annotation of web services: A systematic literature review. *Inf. Softw. Technol.* **2015**, *61*, 16–32. [CrossRef]
55. Elhadad, M.; Badran, K.M.; Salama, G. A novel approach for ontology-based dimensionality reduction for web text document classification. In Proceedings of the 16th IEEE/ACIS International Conference on Computer and Information Science (ICIS 2017), Wuhan, China, 24–26 May 2017; pp. 373–378.
56. Princeton Univeristy. WordNet-A Lexical Database for English. Available online: https://wordnet.princeton.edu/ (accessed on 16 June 2021).
57. Vicient, C.; Sanchez, D.; Moreno, A. An automatic approach for ontology-based feature extraction from heterogeneous textual resources. *Eng. Appl. Artif. Intell.* **2013**, *26*, 1092–1106. [CrossRef]
58. Apache Software Foundation. Apache Open NLP. 2004. Available online: https://opennlp.apache.org/ (accessed on 16 June 2021).
59. Wang, B.B.; McKay, R.I.; Abbass, H.A.; Barlow, M. Learning text classifier using the domain concept hierarchy. In Proceedings of the IEEE International Conference on Communications, Circuits and Systems and West Sino Expositions Proceedings, Chengdu, China, 29 June–1 July 2002; Volume 2, pp. 1230–1234.
60. Russell, S.; Norvig, P. *Artificial Intelligence: A Modern Approach*, 3rd ed.; Prentice Hall Press: Hoboken, NY, USA, 2009.
61. US National Library of Medicine. Unified Medical Language System. 1986. Available online: https://www.nlm.nih.gov/research/umls/ (accessed on 16 June 2021).

62. Khan, A.; Baharudin, B.; Khan, K. Semantic Based Features Selection and Weighting Method for Text Classification. In Proceedings of the International Symposium on Information Technology, Kuala Lumpur, Malaysia, 15–17 June 2010.
63. Yap, I.; Loh, H. T.; Shen, L.; Liu, Y. Topic Detection Using MFSs. *LNAI* **2006**, *4031*, 342–352.
64. Abdollahi, M.; Gao, X.; Mei, Y.; Ghosh, S.; Li, J. An ontology-based two-stage approach to medical text classification with feature selection by particle swarm optimization. In Proceedings of the IEEE Congress on Evolutionary Computation (CEC), Wellington, New Zealand, 10–13 June 2019; pp. 119–126.
65. Kennedy, J.; Eberhart, R.C. *Swarm Intelligence*; Morgan Kaufmann: London, UK, 2001.

66. Lu, S.; Ye, Y.; Tsui, R.; Su, H.; Rexit, R.; Wesaratchakit, S.; Liu, X.; Hwa, R. Domain ontology-based feature reduction for high dimensional drug data and its application to 30-day heart failure readmission prediction. In Proceedings of the 9th IEEE International Conference on Collaborative Computing: Networking, Applications and Worksharing, Austin, TX, USA, 20–23 October 2013.
67. US National Library of Medicine. RxNorm. 2012. Available online: https://www.nlm.nih.gov/research/umls/rxnorm/index.html (accessed on 16 June 2021).
68. U.S. Veterans Health Administration. National Drug File–Reference Terminology (NDF-RT) Documentation. Available online: https://evs.nci.nih.gov/ftp1/NDF-RT (accessed on 16 June 2021).
69. Barhamgi, M.; Masmoudi, A. Lara-Cabrera, R.; Camacho, D. Social networks data analysis with semantics: Application to the radicalization problem. *J. Ambient. Intell. Humaniz. Comput.* **2018**. [CrossRef]
70. Kerem, C.; Tunga, G. A comprehensive analysis of using semantic information intext categorization. In Proceedings of the IEEE International Symposium on Innovations in Intelligent Systems and Applications (INISTA 2013), Albena, Bulgaria, 19–21 June 2013; pp. 1–5.
71. Fodeh, S.; Punch, B.; Tan, P.N. On ontology-driven document clustering using core semantic features. *Knowl. Inf. Syst.* **2011**, *28*, 395–421. [CrossRef]
72. Garla, V.N.; Brandt, C. Ontology-guided feature engineering for clinical text classification. *J. Biomed. Inform.* **2012**, *45*, 992–998. [CrossRef]
73. Lin, D. Automatic retrieval and Clustering of Similar Words. In Proceedings of the 17th International Conference on Computational Linguistics, Morristown, NJ, USA, 10–14 August 1998; pp. 768–774.
74. Qazia, A.; Goudar, R.H. An Ontology-based Term Weighting Technique for Web Document Categorization. *Procedia Comput. Sci.* **2018**, *133*, 75–81. [CrossRef]
75. Rujiang, B.; Junhua, L. Improving Documents Classification with Semantic Features. In Proceedings of the 2nd International Symposium on Electronic Commerce and Security, Nanchang, China, 22–24 May 2009; pp. 640–643.
76. Jena Ontology API. Available online: https://jena.apache.org/documentation/ontology/ (accessed on 16 June 2021).
77. Shein, K.P.P.; Nyunt, T.T.S. Sentiment Classification based on Ontology and SVM Classifier. In the Proceedings of the International Conference on Communication Software and Networks, Singapore, 26–28 February 2010; pp. 169–172.
78. Kontopoulos, E.; Berberidis, C.; Dergiades, T.; Bassiliades, N. Ontology-based sentiment analysis of twitter posts. *Expert Syst. Appl.* **2013**, *40*, 4065–4074. [CrossRef]
79. Wang, D.; Xu, L.; Younas, A. Social Media Sentiment Analysis Based on Domain Ontology and Semantic Mining. *Lect. Notes Artif. Intell.* **2018**, *10934*, 28–39.
80. Penalver-Martinez, I; Garcia-Sanchez, F.; Valencia-Garcia, R.; Rodriguez-Garcia, M.A.; Moreno, V.; Fraga, A.; Sanchez-Cervantes, J.L. Feature-based opinion mining through ontologies. *Expert Syst. Appl.* **2014**, *41*, 5995–6008. [CrossRef]
81. Zhou, L.; Chaovalit, P. Ontology-Supported Polarity Mining. *J. Am. Soc. Inf. Sci. Technol.* **2008**, *59*, 98–110. [CrossRef]
82. Alfrjani, R.; Osman, T.; Cosma, G. A New Approach to Ontology-Based Semantic Modelling for Opinion Mining. In Proceedings of the 18th International Conference on Computer Modelling and Simulation (UKSim), Cambridge, UK, 6–8 April 2016; pp. 267–272.
83. Siddiqui, S.; Rehman, M.A.; Daudpota, S.M.; Waqas, A. Ontology Driven Feature Engineering for Opinion Mining. *IEEE Access* **2019**, *7*, 67392–67401. [CrossRef]
84. Zhao, L.; Li, C. Ontology Based Opinion Mining for Movie Reviews. *Lect. Notes Artif. Intell.* **2009**, *5914*, 204–214.
85. de Freitas, L.A.; Vieira, R. Ontology-based Feature Level Opinion Mining for Portuguese Reviews. In Proceedings of the 22nd International Conference on World Wide Web, Rio de Janeiro, Brazil, 13–17 May 2013; pp. 367–370.
86. Ali, F.; Kwak, K.-S.; Kim, Y.-G. Opinion mining based on fuzzy domain ontology and Support VectorMachine: A proposal to automate online review classification. *Appl. Soft Comput.* **2016**, *47*, 235–250. [CrossRef]
87. Ali, F.; EI-Sappagh, S.; Khan, P.; Kwak, K.-S. Feature-based Transportation Sentiment Analysis Using Fuzzy Ontology and SentiWordNet. In Proceedings of the International Conference on Information and Communication Technology Convergence (ICTC 2018), Jeju, Korea, 17–19 October 2018; pp. 1350–1355.
88. MO-the Movie Ontology. Available online: http://www.movieontology.org/ (accessed on 16 June 2021).
89. Andrea, E; Fabrizio, S. Determining the semantic orientation of terms through gloss classification. In Proceedings of the 14th ACM International Conference on Information and Knowledge Management, Bremen, Germany, 31 October–5 November 2005.
90. Joachims, T. A probabilistic analysis of the Rocchio algorithm with TFIDF for text categorization. In Proceedings of the 14th International Conference on Machine Learning (ICML-97), Nashville, TN, USA, 8–12 July 1997; pp. 143–151.
91. Di Noia, T.; Magarelli, C.; Maurino, A.; Palmonari, M.; Rula, A. Using Ontology-Based Data Summarization to Develop Semantics-Aware Recommender Systems. *LNCS* **2018**, *10843*, 128–144.
92. Ragone, A.; Tomeo, P.; Magarelli, C.; Di Noia, T.; Palmonari, M.; Maurino, A.; Di Sciascio, E. Schema-summarization in Linked-Data-based feature selection for recommender systems. In Proceedings of the Symposium on Applied Computing (SAC '17), Marrakech, Morocco, 3–7 April 2017; pp. 330–335.
93. Nilashi, M.; Ibrahim, O.; Bagherifard, K. A recommender system based on collaborative filtering using ontology and dimensionality reduction techniques. *Expert Syst. Appl.* **2018**, *92*, 507–520. [CrossRef]
94. Mabkhot, M.M.; Al-Samhan, A.M.; Hidri, L. An ontology-enabled case-based reasoning decision support system for manufacturing process selection. *Adv. Mater. Sci. Eng.* **2019**, *2019*, 2505183. [CrossRef]

95. Eum, K.; Kang, M.; Kim, G.; Park, M.W.; Kim, J.K. Ontology-Based Modeling of Process Selection Knowledge for Machining Feature. *Int. J. Precis. Eng. Manuf.* **2013**, *4*, 1719–1726. [CrossRef]
96. Kang, M.; Kim, G.; Lee, T.; Jung, C.H.; Eum, K.; Park, M.W.; Kim, J.K. Selection and Sequencing of Machining Processes for Prismatic Parts using Process Ontology Model. *Int. J. Precis. Eng. Manuf.* **2016**, *17*, 387–394. [CrossRef]
97. Han, S.; Zhou, Y.; Chen, Y.; Wei, C.; Li, R.; Zhu, B. Ontology-based noise source identification and key feature selection: A case study on tractor cab. *Shock Vib.* **2019**, *2019*, 6572740. [CrossRef]
98. Ma, H.; Zhou, X.; Liu, W.; Niu, Q.; Kong, C. A customizable process planning approach for rotational parts based on multi-level machining features and ontology. *Int. J. Adv. Manuf. Technol.* **2020**, *108*, 647–669. [CrossRef]
99. Belgiu, M.; Tomljenovic, I.; Lampoltshammer, T.; Blaschke,T.; Hofle, B. Ontology-based classification of building types detected from airborne laser scanning data. *Remote Sens.* **2014**, *6*, 1347–1366. [CrossRef]
100. Guan, H.; Yang, H.; Wang, J. An Ontology-based Approach to Security Pattern Selection. *Int. J. Autom. Comput.* **2016**, *13*, 16–182. [CrossRef]
101. SWRL Reference. Available online: https://www.w3.org/Submission/SWRL/ (accessed on 16 June 2021).
102. Guan, H.; Chen, W.R.; Liu, L.; Yang, H.J. Estimating security risk for web applications using security vectors. *J. Comput.* **2012**, *23*, 54–70.
103. Martinez-Rodriguez, J.L.; Hogan, A.; Lopez-Arevalo, I. Information Extraction Meets the Semantic Web: A Survey. *Semant. Web* **2020**, *11*, 255–335. [CrossRef]
104. Janowicz, K.; Yan, B.; Regalia, B.; Zhu, R.; Mai, G. Debiasing Knowledge Graphs: Why Female Presidents are not like Female Popes. In Proceedings of the 17th International Semantic Web Conference (ISWC 2018), Monterey, CA, USA, 8–12 October 2018.
105. Li, J.; Liu, H. Challenges of Feature Selection for Big Data Analytics. *IEEE Intell. Syst.* **2017**, *32*, 9–15. [CrossRef]
106. Wu, X.; Yu, K.; Ding, W.; Wang, H.; Zhu, X. Online feature selection with streaming features. *IEEE Trans. Pattern Anal. Mach. Intell.* **2013**, *35*, 1178–1192.
107. Bolon-Canedo, V.; Sanchez-Marono, N.; Alonso-Betanzos, A. Recent advances and emerging challenges of feature selection in the context of big data. *Knowl. Based Syst.* **2015**, *86*, 33–45. [CrossRef]

MDPI

St. Alban-Anlage 66

4052 Basel

Switzerland

Tel. +41 61 683 77 34

Fax +41 61 302 89 18

www.mdpi.com

Future Internet Editorial Office

E-mail: futureinternet@mdpi.com

www.mdpi.com/journal/futureinternet